DESIGN IN STEEL

 STAND RD

DESIGN IN STEEL

MEL BYARS

Research by

Brice d'Antras

Cinzia Anguissola d'Altoé

Laurence King Publishing

Published in 2003 by Laurence King Publishing
71 Great Russell Street
London WC1B 3BP
United Kingdom
Tel: +44 20 7430 8850
Fax: +44 20 7430 8880
e-mail: enquiries@laurenceking.co.uk
www.laurenceking.co.uk

This book has been made possible through the
sponsorship of Arcelor.

A catalogue record for this book is available from
the British Library.

ISBN 1 85669 313 9

Design: Plan-b, London

Printed in Italy

Frontispiece: Ronen Kadushin, "High Square
Dance" table, 600 x 600 x 750mm. Painted steel.

To Pierre Bourrier

Contents

"Are nails made of butter? Are knives made of cloth?"

Design in Steel is as much a survey of new design and what designers are thinking as it is a book about steel.

There was a time in my books when I felt an apology, or a notice, was necessary to explain the inclusion of the large percentage of Italian designers. The situation still exists and is due to the ongoing healthy climate for design in Italy. With my paranoia rising once again, I felt that I should explain why I included so many designers from Israel in this volume. Yet, on closer inspection, I discovered that there were not as many here as I had initially thought. I was being influenced by the recent efflorescence of good design in that wee place. Much of this has, no doubt, been fostered by the commendable pedagogy offered by the almost-century-old Bezalel Academy of Art and Design there and its current director of industrial design, Ezri Tarazi. And there are certainly other factors, possibly more social than scholastic, contributing to the aesthetic milieu in Israel today.

The Israelis have joined a throng of designers the world over whose nationality can no longer be revealed by the look of their products. Even so, and sadly so, many worthy prototypes, or ideas, in every country (possibly fewer in Italy) are being held prisoner by the lack of enlightened manufacturers and are spending their lifetimes on dusty shelves as lonely one-of-a-kind objects. I have attempted here to keep them to a minimum, including instead mostly in-production merchandise. However, I could not resist including some commendable examples.

To overcome the absence of local manufacturers, some designers, for example from the US, the UK, and France, are seeking out producers or marketers in Canada and Italy; although this is not to say that there is absolutely no local manufacturing in their homelands. Still others, in Britain, have turned to self-production (either by manufacturing the products themselves on their own premises or by hiring fabricators) and self-marketing. Taking on three roles—designer, manufacturer, and marketer—is, however, potentially overwhelming and risks disappointing sales.

There are few success stories like Ernesto Gismondi—designer and entrepreneur—and his Artemide firm. And there are even fewer eagle eyes like those of Alberto Alessi, the artistic director of the wealthy Alessi enterprise, who continues to reach out from the little Italian burg of Crusinallo to some of the more remote talent in the world design community. Both Artemide and Alessi products are included in this volume.

Contributing to the global scope of *Design in Steel* is the work of other Italians and, as mentioned, Israelis, as well as French, American, Argentine, British, Danish, Dutch, Finnish, German, Japanese, Norwegian, Swedish, and Swiss designers. The roster also includes a number of female designers. I, and those who helped me, made no concerted effort to include women. I mention this not to be politically correct but rather to flag women's growing adoption of industrial design as a profession. This brings to mind my embarrassment at having initially thought that steel was a masculine, tough material. This preconception proved to be, as many entries illustrate here, erroneous.

Still, steel is unique. When seeking solutions for certain products, there are often simply no other appropriate substitutes for steel. As Italian designer Enzo Mari is often quoted as asking, "Are nails made of butter? Are knives made of cloth?"

When the designers here were asked what they thought about steel—whether they liked it, why they specified it, how it behaved, and so forth—their responses were intelligent, insightful, lucid, even poetic. One of them, American designer Karim Rashid, simplified his passion thus: "I love steel."

When you take a look at some of the

objects in this book, you might assert, "This isn't about steel. It's about another material." However, you should then ask yourself the question, "Would the construction have been possible, would it have worked, aesthetically or mechanically, without the component of steel?"

All entries have been organized alphabetically by designers' surnames or, where products were created in manufacturers' in-house studios, by the manufacturers' names. Some manufacturers, such as Swatch, do not wish their designers to be known. Others, such as Simplicitas and Rapsel, are happy to reveal their designers but are more interesting as ensembles and therefore appear under their own names. Alessi and Cappellini products appear under the individual designers' names. Such ordering by designer or manufacturer name is intended to make for a more interesting read than product-specific chapters and allows for the full scope of a designer's or firm's oeuvre to be seen in one place.

Concerning how *Design in Steel* came into being, the world's largest producer of steel, Arcelor, generously commissioned it. François Barré, a consultant to Arcelor and someone who worked at the French Ministry of Culture for many years, proposed a book on the domestic uses of steel, one that would reveal the ubiquity of this often unnoticed material in everyday life. The firm and its former director of communications, Pierre Bourrier, gave me free rein, which meant that I was solely responsible for choosing the final entries. No designer or manufacturer paid to be included; no preferential treatment was given except to work that best served the subject.

And concerning books themselves, my late mother, who never fully understood just what it is I do for a living, thought, like so many others, that books happen by magic,

that they are simply written, printed, and then sold in stores. *Voilà.* The actual writing is the least complicated part of the process. There is the task of gathering the images, the documentation, and the research that consumes me and particularly the intelligent people—Cinzia Anguissola d'Altoé and Brice d'Antras—who help me indefatigably. Just why they do it, book after book, eludes me. La sig.ra Anguissola is a busy, practicing Milanese architect with a big family and Monsieur d'Antras is the new director of the Institut Français in Hanover. Could I do without them? Probably not.

La sig.ra Anguissola also helped me with the Italian translations of the designers' comments. I struggled with other languages, too, and edited some of those statements provided by designers or manufacturers whose English needed a little polishing. Regardless, translations never serve the original. Indeed, while Italian can be lyrical, English is often not, at least in discussions about aesthetics and technology. So I beg the designers and manufacturers to exert great patience when they see how I might have failed to capture their subtleties and charming sincerity. I thank them for sharing their thoughts—even intimate ones, in some cases. The manufacturers were extremely helpful in acquiring some of the designers' comments, because most of the designers are freelance and work elsewhere.

Neglecting to acknowledge the publishing staff would be egregious. Laurence King, the head of the firm; Jo Lightfoot, the senior commissioning editor of design and graphics; Liz Faber, the managing editor; Laura Willis, the public relations person; and Nell Webb, the senior editor, met my incessant whining and bad behavior with good humor and nurturing support.

At Arcelor, Valérie Dusséqué of the marketing department offered her depth of

knowledge in the verification or correction of the techniques and terminology discussed. And Nicole Eleouet made my relationship with the firm very easy.

No doubt I am omitting to acknowledge the help of others. The suggestions of Marianne Russell of the Arango store in Miami and Madame Levitte in Paris, who was undauntingly helpful in so many matters, resulted in the book's having more depth than might otherwise be the case.

—Mel Byars

DESIGN IN STEEL

Jerker Andersson
"Abisko" flask

As any laborer who has carried a flask in his lunch pail to keep a beverage hot for a long time will confirm, the lining of traditional vacuum flasks easily breaks when dropped. And, in the normal course of use, they are often dropped. Breakage results because the inner vessel is made of thin glass with a vacuum space between the double walls. A Swedish housewares manufacturer has diverted from its normal plastics inventory to offer a solution, designed by Jerker Andersson of Krypton Form AB in Göteborg. Its all-metal flasks, made of 18/8 stainless steel, will not break or leak and offer excellent heat retention. There is also no need to remove the screw stopper when pouring. Sales of more than 100,000 pieces over the past two years attest to how well the idea has been received.

| "Abisko" one-liter flask. Available in four sizes, the largest (324mm high), shown here, retains 100°C liquids up to 72°C after 6 hrs or up to 60°C after 12 hrs.

date of design | 1998, produced since
manufacturer | Hammarplast AB, Tingsryd, Sweden

---

designer says | "At the outset of this project I bore in mind the words 'steel vacuum flask.' I felt that the design should be linked to the shape of a flask and have an appealing appearance. The top and bottom should be easily distinguishable. The diameter at the bottom is wider, and the flask therefore stands firmly without toppling over. At first, it was difficult to accomplish such a design, but we solved this by fitting the inner container from beneath. The body of the vacuum flask is made of lightly brushed stainless steel. I think these flasks serve equally well at the dining table as out in the countryside."

| The cap (left) also serves as a drinking cup. It fits tightly over the gray injection-molded polypropylene dividing stopper inside (below). And, when replaced, it automatically presses down the red button (below opened) that seals the stopper to eliminate leakage.

Ron Arad

Furniture and accessories

After moving from Israel to London 30 years ago and graduating from the Architectural Association there, Ron Arad established the One Off studio with Caroline Thorman. He has since become one of the world's top designers and Israel's best-known son in the design world. Even though the rather dramatic chair shown here was intended for limited production, much of Arad's current work is serially produced—for firms such as Alessi—in very large numbers.

facing | "Narrow Pappardelle" (chair: 400 x 1050mm; 1000mm rolled; 4000mm unrolled) is polished stainless-steel woven links with stainless-steel profiles on chair sides, assembled by TIG welding. (TIG welding, or tungsten inert gas welding, is also known as GTAW, or gas tungsten arc welding.) This woven-steel product is used in the food industry.

date of design | 1992, produced 1993–2000
manufacturer | Ron Arad Associates Ltd, London, UK

---

right top and middle | "Hanger" towel hook/clothes hanger (285 x 285mm, bent forward, or 223mm deep folded back) is made of rods and a solid wall-mounted base, both in stainless steel. The hanger element can be upright, inverted, or flat; or, with the inner pieces, bent forward.

date of design | 1995, produced since
manufacturer | virtuallydesign.com, Monza, Italy

---

bottom | "Baby Boop" steel dish is available as a two-section (230 x 195 x 40mm), three-section (230 x 200 x 40mm), or four-section (290 x 210 x 40mm) container. Formed by cold-pressing (blanking and punching machines) and surface-finished by other machinery.

date of design | 2000, produced since
manufacturer | Alessi S.p.A., Crusinallo di Omegna (VB), Italy

---

designer says | "When I started 'designing' in steel, I used this material because I had to make everything myself. I was not a craftsman. I always thought that for wood you needed to be a craftsman and for plastic you needed the industry. On the other hand, steel is a very forgiving material; you can cut it, weld it, bend, hammer, polish it, etc. With steel you can also change your mind, make mistakes, and correct them; you can improvise. If you look at the early pieces I have made in steel, they look very primitive and amateurish and that is exactly what they were. In time, my team and I got a lot better. We actually made pieces in steel that we believed no one else could make. Of course, we were wrong! The minute I realized I was in danger of becoming a craftsman, I closed the metal workshop and moved on. But I still enjoy having had the experience; it does help when designing for the industry."

Photographs courtesy virutallydesign.com ("Hanger"), Alessi S.p.A. ("Baby Boop") and Wilhelm Moser ("Narrow Pappardelle")

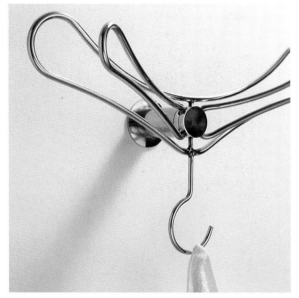

Shin and Tomoko Azumi
Speakers and furniture

Formerly active in Japan, the Azumis established a studio in 1995 in London, shortly after they graduated together from the Royal College of Art. The attentive approach of this husband-and-wife team is evident from the examples shown here and has been recognized by a number of manufacturers in Europe as well as in Japan.

facing | "H2" speakers (111 x dia. 288mm) are available in a light color or black. The black model (with the screen removed) reveals the inner workings—woofer (large in the center), tweeter (small in the center below), and the sound tunnel (left of the woofer).

below | "H1" (right) and "H2" (left) directional speakers. On the rectangular "H1," the sound tunnel is placed longitudinally. The screen, in this case, is not a traditional fabric but rather a perforated and pressed steel that has been powder-coated to match the plastic case.

date of design | 1999, produced since
manufacturer | TOA Corporation, Kobe, Japan

designers say | "Perforated metal was used, firstly because it facilitates a clear sound quality, secondly because it broadened the possibilities of color and shape in a cost-effective way, and thirdly because it offers easy maintenance."

Photography courtesy TOA Corporation

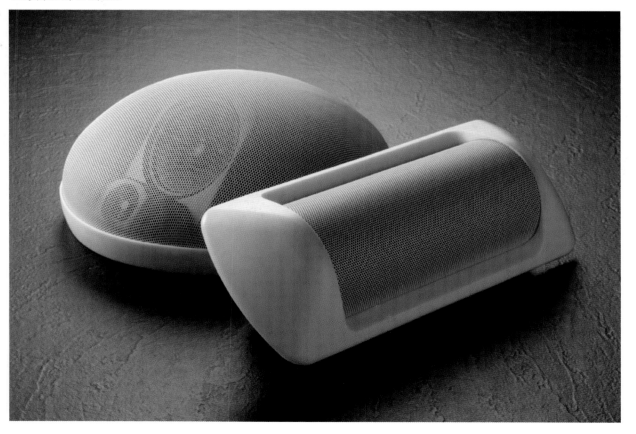

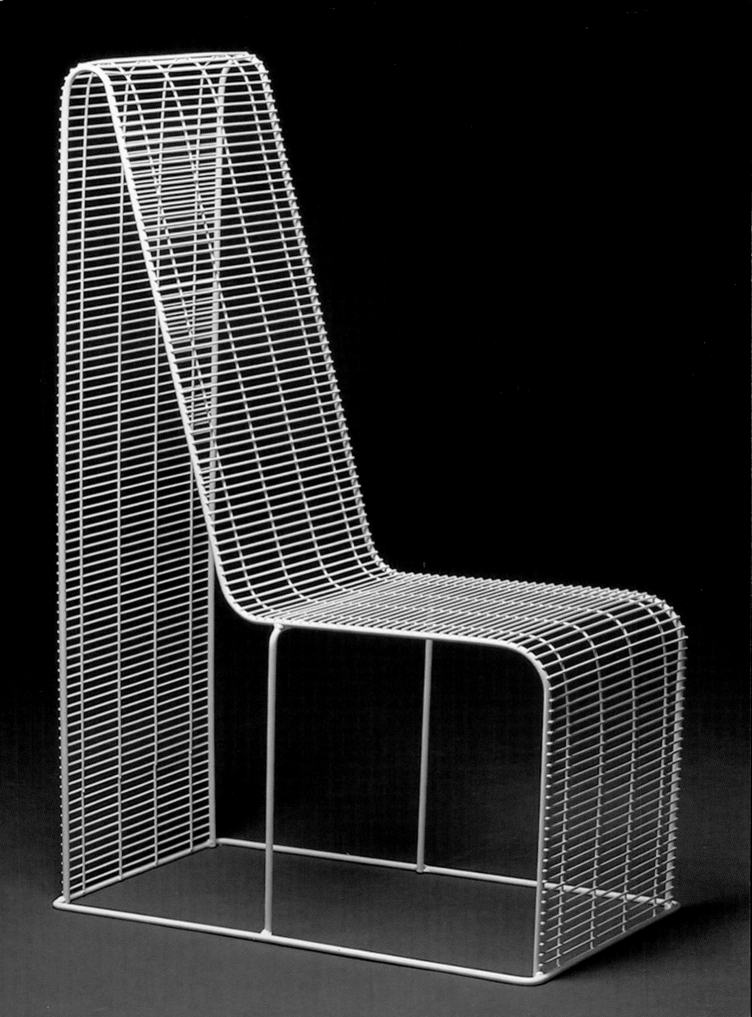

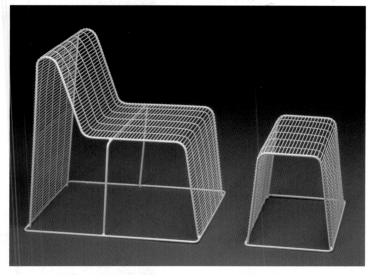

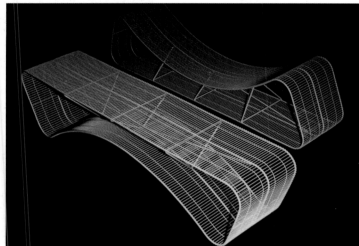

facing | "Wire Frame" high-back chair (360 x 703 x 1100mm), as well as other models in the "Wire Frame" range employ zinc-plated, spot-welded, powder-coated mild-steel rods. The series is produced by James Gott through his firm Meshman; Gott is a designer in his own right.

top | "Wire Frame" chair (550 x 770 x 730mm) and stool (500 x 320 x 380mm).

middle | "Wire Frame" reversible bench (1700 x 360 x 400mm).

bottom | "Wire Frame" stool (384 x 250 x 380mm). When a group of stools is configured sidewise, a shelving system, or étagère, becomes possible.

date of designs | 1998–99, produced since
manufacturer | Meshman, Manchester, UK.

—————

designers say | "We used a specific 1:6 grid (½in x 3in). This rectangular proportion successfully creates a visual moiré effect when it is seen through overlapping planes. After we designed this series we made an installation of wire-frame furniture and 1,500 bungee cords that maximized the moiré effect."

Photography by Julian Hawkins

Mario and Claudio Bellini
"Palmhouse" whistling kettle

One might legitimately ask, "Is it possible to reinvent the kitchen kettle?" The answer is, "Evidently so," since accomplished designers Mario Bellini and his son Claudio of Milan have come up with a new solution for this age-old appliance for boiling water. Somewhat unusually shaped, the "Palmhouse" is made of 18/10 AISI 304 stainless steel and offers handles and spout in a choice of colors. The construction is sophisticated.

facing | The 1.7-liter-capacity kettle features a three-layered bottom (steel/copper/steel). The steel body, satin-finished or polished, is stuffed and pressed by computer-numeric-controlled machines (CNC), run by PLC. The polyamide handle and spout are injection-molded in a warm chamber to make the surface smooth. Metal parts are TIG welded (see p. 13). A ten-position machine tool, moving on five controlled axes, polishes the surface.

right | A prototype in wood was built prior to production.

below left | Sketches were rendered by Mario or Claudio Bellini during conception. The example here is no doubt one of many.

below right | The handle and spout are available in five colors—white, dark gray, sky blue, light green, and wine red.

date of design | 2000, produced since
manufacturer | Barazzoni, Invorio (NO), Italy, for Cherry Terrace Inc., Tokyo, Japan

Photographs by Leo Torri

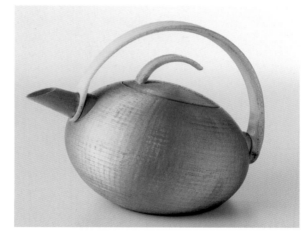

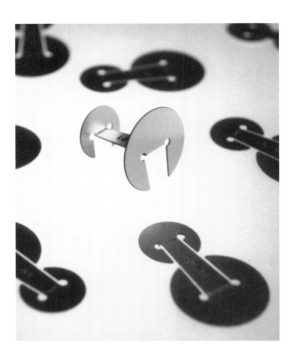

Sebastian Bergne
Kitchenware, lighting, and jewelry

While many designers like to play and experiment, expressing their sense of humor, most of them will embrace an opportunity to design for a significant, albeit serious, client. London designer Sebastian Bergne takes a jovial approach that calls on the use of metals—illustrated here by his manipulation of refuse, of a memory-retaining metal sheet, and of sprung stainless steel for make-do jewelry. However, his breakfastware for a major manufacturer that specializes in stainless steel assumes a more serious approach.

above | "Emergency" cufflinks (30 x 18mm) pretty much reveal, if you know the name, their story. Of those who still wear them, who hasn't misplaced their cufflinks? These offer a cheap and accessible solution to the crisis. Also, they can be worn as regular links with no apologies. Made from a 0.15mm-thick sheet of stainless steel, they are cut out by acid-etching.

date of design | 1993, produced from 1994
manufacturer | The designer

designer says | "Sprung stainless steel gives the necessary stiffness and strength, even in a very thin sheet, to allow for folding along half-etched lines without excessive weakening of the material."

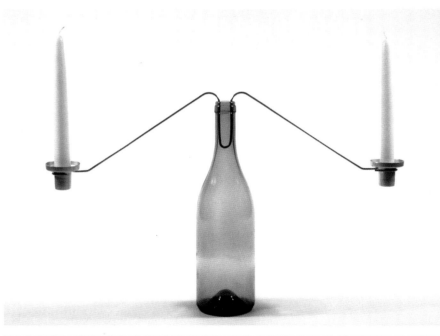

left | "Candloop" candelabra (470 x 106mm) requires a host or hostess to add the element of an empty wine bottle. The 3mm-diameter stainless-steel bent wire is simply inserted into the mouth. The candles are held by a metal of another kind. The wire is shaped by computer-numeric-controlled machinery.

date of design | 1998, produced from 1999
manufacturer | Wireworks UK Ltd, London, UK

designer says | "Stainless-steel wire is the perfect material for this product. Not only does it have the right physical characteristics to achieve the spring action required for retention in the neck of the wine bottle, but it also has the right composition, color, material, and aesthetic values."

left | "Lampshade 1" and "Lampshade 2" (125 x 360mm) are acid-etched from a single 0.15mm-thick stainless-steel sheet. A simple solution, the shades, depending on the model, can be slipped over a hanging bulb as a downlighter, or under it as an uplighter.

date of design | 1990, produced from 1991
manufacturer | The designer; subsequently Radius GmbH, Brühl, Germany

designer says | "The flexibility and spring quality of the material allows the lampshade to be clipped onto the bulb or fitting. The action of attaching the shade transforms the object from a flat sheet to a curved reflector. When not in use or packaged for transport, the shade returns to its flat state."

| "Kult" breakfast set marries metal with pressed glass. The Cromargan© 18/10 stainless steel is shaped with automatic presses and then brush-finished. (Cromargan is WMF's proprietary 18/10 stainless steel. "18" indicates the percentage of chrome which makes the material stainless; "10" indicates the percentage of nickel which makes it even more stainless.) The group comprises an egg cup, creamer, sugar bowl, jam jar, butter dish, and toast rack.

right | The egg cup includes a container to be used either as a stand for the egg holder or as an egg-shell refuse bin.

middle | Containers for cream, jam, and sugar.

bottom | The toast rack is designed to hold both toast and, when the lid is removed, untoasted bread.

date of design | 1997, produced from 2000
manufacturer | WMF A.G., Geislingen/Steige, Germany

designer says | "For me, a project with WMF meant primarily stainless steel. They are one of *the* reference manufacturers in this material. Recognizing a company's culture and strengths is an important part of the designer's job. Stainless steel is ideally suited to this breakfast-set application because it is a high-quality, hard-wearing, unbreakable material with a patina that improves with age."

Jeffrey Bernett

furniture

The American designer Jeffrey Bernett has, like others in the US and Europe, become active in designing for firms in Italy whose manufacturers have begun to recognize foreign talent. The chairs shown here are part of a collection that includes a bed and a side table. (For another Cappellini product, see p. 32.)

top | "Monza" low-version chair (650 x 550 x 600mm) is built from cold-rolled steel sheets. Rolling and shearing machines and drilling and threading tools form the shapes, after which they are powder-coated. Mounting screws are inserted in the threaded holes in the vertical plane to hold the seat.

bottom | "Monza" high-version chair (530 x 500 x 700mm) is similarly constructed to the low-version model.

date of design | 1997, produced from 1999
manufacturer | Cappellini, Arosio (CO), Italy

designer says | "In discussions with the manufacturer about this project's parameters as far as the materials and manufacturing capabilities were concerned, we determined that cold-rolled steel would be the best choice of material since the design could be easily realized by using rolling and bending machines. The manufacturer had already successfully used this equipment on previous projects. Since my design was fairly straight-forward, there was essentially no new investment required for tooling."

Bertocci Studio
Bathroom sink

One of the more simple bathroom sink designs by this manufacturer, the "144–140" model shown here attests the favorable response by the public, designers, and manufacturers to the use of glass bowls. The glass, of course, visually outshines the essential elements which are made of stainless steel. The basin—available in white, aquamarine, or cobalt-blue glass—is supported by the metal wall-mount shelves and the flow drain. Unlike those made of cast iron or PVC, the underpipes and stopper lever are attractive enough to be exposed. While established designers have been commissioned by Bertocci to create many of their bathroom sinks and accessories, the model here was designed by the in-house staff.

| The center-positioned wall-attached section of the "144–140" model accommodates the mixer. This section and the flow-drain fitting firmly hold the basin in place with soft-plastic washers. A direct metal-to-glass connection is undesirable due to possible glass damage and water leakage. The separate wall-attached shelves and the mixer support are made of polished high-thickness stainless steel that has been laser-cut and bent. The four pieces of the wall support sections are bored with 350mm holes, suitable for installing single-lever mixers made by this manufacturer or by another. However, the mixer here offers color coordination.

date of design | 1999, produced from 2000
manufacturer | Bertocci Arnolfo e figli S.r.l., Sesto Fiorentino (FI), Italy

Photography by Mario Corsi

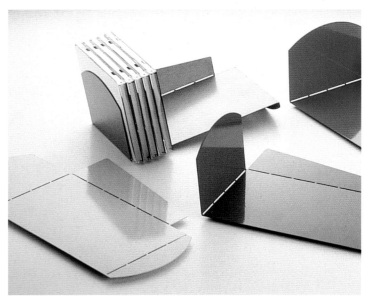

Blu Dot
Furniture and office accessories

Designers who have established national or international reputations but who work outside metropolises such as New York, London, Paris, and Milan are becoming more commonplace. The Blu Dot team, comprising John Christakos, Maurice Blanks, and Charlie Lazor—close friends since university—is an example of this localism. From the Northeastern seaboard of the US, Christakos and Lazor moved to Minneapolis and Blanks to Chicago. When they formed a partnership in 1997, they could not find a firm that both produced reasonably priced furniture and liked their approach. Blu Dot therefore had its own designs made by contractors, although it now works for large firms such as the Target stores.

facing | "2D:3D" desktop accessories are 22-gauge (0.9mm) cold-rolled steel, laser-cut by computer-numeric-controlled (CNC) machinery and powder-coated. For home or office, the system is inexpensive and practical and designed to be fun. The collection includes an in/out letter box (289 x 250 x 125mm). Purchased as flat pieces, end-users are expected to bend the objects into shape—hence the name "2D:3D," or two dimensions into three dimensions. The steel is perforated and painted steel blue, burnt orange, mint green, white, or with a brushed natural finish. (For a similar concept, see pp. 102–103.)

top | Close-up view of the "2D:3D" system and the perforations that permit precise bending.

middle | Letter holder (250 x 100 x 175mm).

bottom | Desktop CD holder (225 x 125 x 100mm).

date of design | 1999, produced since
manufacturer | Blu Dot, Minneapolis, Minn., US

designer says | "A few years ago, we were working on some fairly complicated bent-sheet products and trying to figure out how to keep the two-dimensional sheet flat and to let the consumer do the three-dimensional bending operation. Plastic was not rigid enough; aluminum proved too fatiguing. We discovered that we could do the seemingly impossible: to hand-bend steel by laser-cutting perforated lines into it, thereby making it weak enough to bend by only the pressure of the hand but strong enough to hold together as a rigid form. We are now looking at this perforating strategy for furniture products because it wins big in the cost and flat-packing categories and makes 'assembly' an interactive event. What fun it would be to fold your steel furniture together on your living room floor without any tools!" —Charles Lazor

Photography by Charlie Lazor (facing, left top, and left middle) and Parker Photographic (left bottom)

facing and below left | "Lil' Buddy" computer desk (850 x 500 x 775mm). The body is made with 14-gauge (2.0mm) cold-rolled steel that is laser-cut and brake-pressed by computer-numeric-controlled (CNC) machinery and powder-coated.

below right | The legs are mounted to the steel "box" through no-tool sockets and knurl-nut connections.

date of design | 2001, produced from 2002
manufacturer | Blu Dot, Minneapolis, Minn., US

Photography by Andrew Balster

designer says | "Blu Dot is interested in 'affordable design'. We look to common, industrially made materials and low-cost industrial methods to produce forms—typical component parts—that can be assembled into furniture. Steel is always in front of us because the economics are incredible. It is very cheap, yet it has unassailable integrity. Steel is fabricated in such a direct and primary manner: sheared, bent, stamped. We try to hold onto that essential 'operation of making' so it becomes self-evident in the finished piece, and steel performs so well in this regard. Bent steel has the properties of both desirable surface and structure at the same time. It is fairly unique in this regard and the possibility of creating both a functional, programmable surface and a structure in a single stroke is delightfully efficient to us."
—Charles Lazor

Constantin and Laurene Leon Boym
Kitchenware and furniture

facing | "Tin Man" kitchen canisters (254 x dia. 178mm) by Constantin Boym. The design of these canisters called upon a traditional object—the coffee can. However, stainless steel, rather than tin, is used and has been deep-drawn and polished. Lathe machinery formed the grooves.

below left | Prototypes use real coffee cans.

below right | Detailed drawings (an elevation-and-plan view shown) assure final-production accuracy.

date of design | 1990, produced from 1994
manufacturer | Alessi S.p.A., Crusinallo di Omegna (VB), Italy

designer says | "I have always seen coffee cans as beautiful objects, in which function, aesthetics, and tactility are in perfect balance. Back in 1990, I envisioned a whole set of objects, based on the coffee-can theme. It took four years to convince Alberto Alessi to issue part of the collection. Polished stainless steel was the only logical choice for the material and for Alessi's expertise. I origi-nally envisioned a steel wire for the handle, which was later replaced by a more conventional Alessi-style wooden knob." –Constantin Boym

Russian émigré Constantin Boym and his American wife and studio partner Laurene Leon Boym admire the commonplace and reinvent traditional objects through an acute insight. Boym established his eponymous studio in 1986; his wife has been a member since 1995, and the two work both separately and together. Boym has become known for his "Souvenirs for the End of the Century," a unique take on miniatures of historical monuments and buildings. Rather than focussing solely on new forms and materials, the Boyms' work is deeply rooted in the history and meaning of objects.

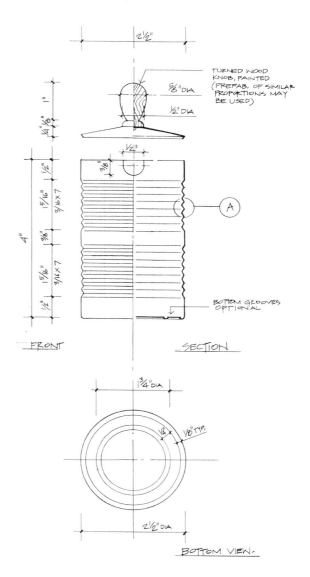

top | "Paper Table" (508 x 390 x 330mm) by Constantin Boym is so named because of the inclusion of a paper pad and pencil that are standard equipment and located in the well of the top. The final product makes use of a mundane material—galvanized, bent, and riveted sheet steel. Galvanizing is produced when steel is immersed in a bath of molten zinc, thus applying a coating of an iron-zinc alloy.

bottom | Prototype in foam core (a paper-covered foam board).

date of design | 1998, produced since
manufacturer | DMD, Voorburg, The Netherlands

———

designer says | "Originally intended as a telephone table, a pad of paper and pencil were incorporated into the design. However, it is possible to use it alternatively, for example when the paper runs out. Steel was chosen because I wanted a connotation with equipment, not furniture, for example warehouse shelving or restaurant-kitchen equipment. Fortunately, DMD [the production entity of the Droog Design group] already had a supplier that could manufacture galvanized-sheet objects." –Constantin Boym

Top photograph courtesy DMD

left | "L.I.M." shelf (406 x 406 x 178mm) by Constantin and Laurene Leon Boym. The name "L.I.M." is an abbreviation of, as well as a play on, Ludwig Mies van der Rohe's mantra, "Less is more." The shelf is made of a powder-coated bent steel sheet.

below | Illustrations offer suggestions for horizontal or vertical use as multiples.

date of design | 1999, produced since
manufacturer | Pure Design Ltd., Edmonton, Alberta, Canada

---

designer says | "The 'L.I.M.' shelf was specifically designed as a collector's shelf to hold souvenirs, pottery, any possible collectibles which most people have in their homes. In this sense, the name (with a reference to Mies) is ironic: the shelf itself is quite minimal, but it is only complete when 'stuff' fills it up. Steel was chosen because powder-coated sheet steel was the material of expertise of the manufacturing company. In this case, it was an ideal choice anyway because of the abstract qualities that powder-coated sheet steel possesses. A reference to Donald Judd's sculptures came as a bonus."
—Constantin Boym

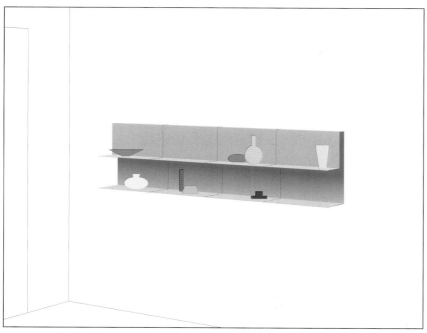

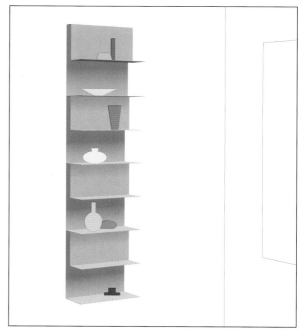

Stephen Burks/Readymade Projects
"Display" shelving system

Stephen Burks, working within his multi-disciplinary studio Readymade, has developed ideas, products, furniture, and furnishings for a number of international clients, including for his own Readymade brand. The shelving shown here, for Cappellini, serves as an example of how refinement and workability can be best achieved when a designer works closely with a manufacturer—in this case Burks with Giulio Cappellini and his staff. (For another Cappellini design, see p. 22.)

top | The understructure reveals the cross bar connection. The material is 18-gauge (1.3mm) powder-coated, bent steel sheeting for the shelves, stainless-steel cross bars for the bracing, and nylon feet. A stainless-steel pin holds the cross bars at their intersections. Mild-steel Allen-head bolts hold the cross bars to the underside of the shelves.

middle and bottom | The long three-tier version (2135 x 350 x 1100mm). Available in orange-red (middle), yellow-green (top), brown, white (bottom), and brushed stainless steel.

date of design | 2000, produced since
manufacturer | Cappellini S.p.A., Arosio (CO), Italy

designer says | " 'Display' was derived from my ongoing attempts at alternative structural methods for furniture. The goal was to make a light, formally direct, open shelving system that would be accessible from both sides with minimal structure. Steel was chosen both out of my own personal desire to explore the material and the qualities of strength and lightness that are possible with the material. I also wondered if it were possible to domesticate the material elegantly within the typology of shelving and break the paradigm of steel shelving being so commonly expressed industrially in the home and office."

Ricardo Bustos
Furniture

Ricardo Bustos is an Argentine working and living in Paris. The tables shown here are composed of intricately folded flat steel. They are available in numerous other forms, in bright or subdued colors, and are suitable for various rooms or uses.

designer says | "Initially, I designed lighting fixtures by folding paper. Then, when I began designing furniture, I continued using the same method. Steel was the obvious material that came to mind for furniture because of its structural properties. And simple bending and welding made it possible for me to have my pure forms and flat profiles realized."

above | Double table (400 x 600 x 4010mm). 4mm-thick steel sheets are formed by cutting and folding machinery, like the table below. The surface is hand-brushed.

below | The dining table features double-plane legs and a double-top strut (as in the model above) that have been multiply folded to form a space between the front and back planes of the legs as well as at the top where a wooden slab can be inserted. The wood-metal geometry offers a rigidity that eliminates the need for screws or glue.

date of design | 2001, produced since
manufacturer | Ricardo Bustos Design, Paris, France

Casimir
"Fancy Table No. 2"

The designer Casimir works from a studio and small show-room with production facilities in Belgium. The table here began as "Fancy Table No. 1," a version in solid wood. The shape of the metal version clearly recalls the table's origins, almost as if the 1991 wood model were a prototype. The designer, whom *Wallpaper** magazine named "the international man of mystery," works closely with Wouter Cottyn, an old friend whom he met while studying design at the academy in Genk and a specialist in metalworking. It is Cottyn who is responsible for the steel table's construction.

| "Fancy Table No. 2" (1200 x 300mm) is composed of two cold-formed (or machine-pressed), welded, stainless-steel forms. The surface is brushed.

date of design | 2000, produced on request
manufacturer | Casimir Meublen MV, Heusden-Zolder (Limberg), Belgium

———

designer says | "About 11 years ago we started production of the [wooden] 'Fancy Table (Schaal) No. 1.' To use the 'Fancy Table' outside, we chose to make it in stainless steel (or inox). In addition to inox's suitable outdoor use, it accommodates a large size."

Didier Chaudanson
Lighting

The conception behind these lighting fixtures by Didier Chaudanson, a designer working in Paris, was made possible by laser machinery and bulb heat. The former enabled the images to be cut out on the metal diffuser; the heat of the bulb, assisted by a loose top-center bolt, makes the diffuser rotate. Thus, on one version, a man can be seen trotting along, and, on another, the lines of a message become readable.

facing | "Lumen" sconce (120 x 150mm) or table lamp (250 x 90mm, not shown) includes a stainless-steel shade and base. The negative images are laser-cut in the diffuser which is stamped and cut out from the metal plate, either mirror- or brush-finished. Epoxy glue and riveting joints complete the construction.

top and bottom | "Chrono" sconce (120 x 150mm) or table lamp (250 x 90mm) is produced by the same methods as the "Lumen" (facing) and includes the same features.

date of design | 2000, produced 2000–2001
manufacturer | The designer

designer says | "The use of stainless steel, whether brushed or polished, means that the metal can remain unfinished, which thereby eliminates the necessity of a finishing stage that might produce disappointing results."

Photography by Didier Chaudanson

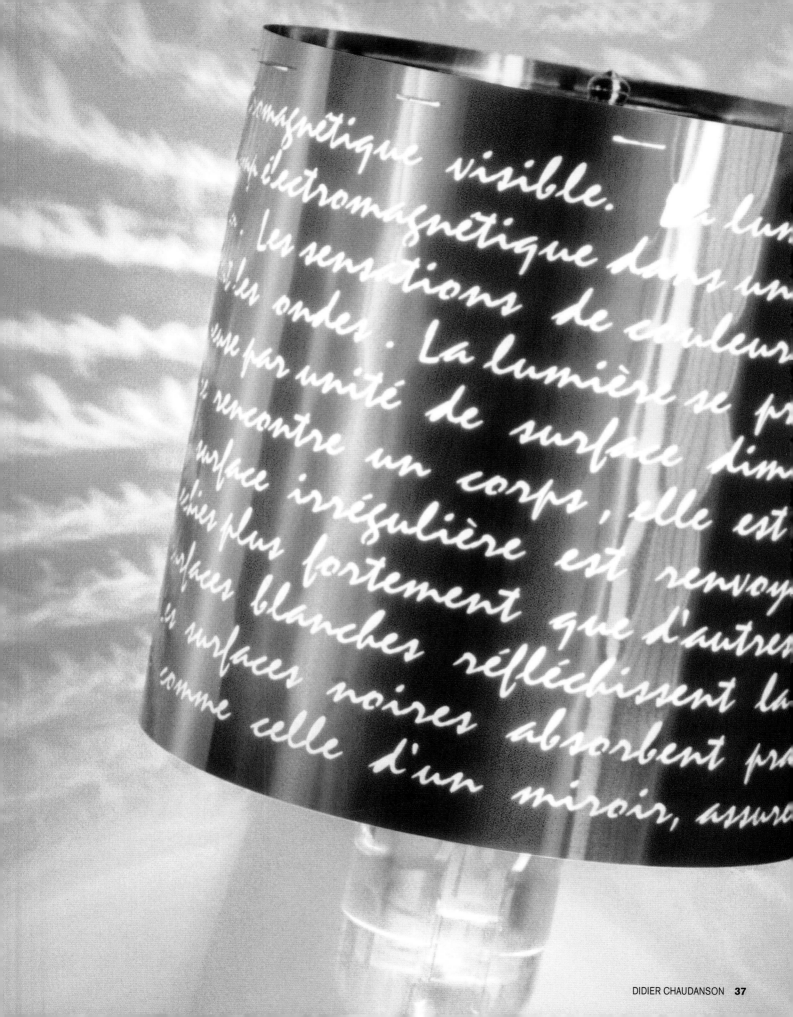

Antonio Citterio with Oliver Löw

"Glossy" table

A prolific designer, Antonio Citterio works from his large studio in Milan, sometimes with Oliver Löw and sometimes with his American-born wife Terry Dwan. Just what Löw contributes to Citterio's work is not known. Citterio and Löw have designed a number of successful furniture models for the Kartell firm. The manufacturer has sold, for example, 230,000 pieces of the duo's three-drawer "Mobil 2020" cabinet in PMMA (paramethoxymethylamphetamine) and a steel frame. The result of thorough and intelligent marketing strategies, new Kartell designs must endure three or more years of developmental scrutiny prior to final production.

facing and below | "Glossy" table is offered with a square (1300 x 720 x 1300mm), round (720 x dia. 1300mm), or ovoid (1200 x 720 x 2000mm) top. While the thin, pristine top is distinctive due to the sparkling, high-gloss polyester-lacquered finish, the spindly chromium-plated and welded, tubular steel leg structure completes the dynamic image. (Also available with a laminate top.)

date of design | 2000, produced from 2001
manufacturer | Kartell S.p.A., Noviglio (MI), Italy

Erik Demmer and Christian Schäffler
Bathroom fittings

A configuration of intersection cylinders, the taps by German designers Erik Demmer and Christian Schäffler feature perhaps the narrowest diameter base (36mm) available from any manufacturer. The modular design system allows various model configurations. Different outlet lengths and adjustable faucet heights make for several uses and installations. The "Pollux" bowl, named after a star in the universe, is available in stainless steel or one of several treatments of glass and in two models. The bowl has received prestigious design awards. (Vola, the manufacturer, is the firm that began producing Arne Jacobsen's innovative in-wall tap from the early 1960s.)

facing and left bottom | "Marathon II" faucet (with a lever or separate hot/cold handles) is available with 133 or 210mm spouts and 133, 185, or 333mm stems. The finish resists corrosion and damage from cleaning agents. The unit is all stainless steel and can be wall-mounted. Accessories include hand shower extensions, a flexible hose, and a soap dispenser.

left bottom | "Marathon II" armature.

left top | "Pollux 2" (dia. 355mm) bowl is a double-wall stainless-steel model, brushed or polished and, inside the walls, filled with sand for noise reduction. The wall version (shown) is recommended by the manufacturer but requires specially placed behind-the-wall plumbing.

date of designs | "Marathon II" faucet and armature: 1999, produced from 2001; "Pollux 2" bowl: 1998, produced from 1999
manufacturer | HighTech + Vola A.G., Munich, Germany

do
Furniture and lighting

The Droog Design group was founded by Gijs Bakker and Renny Ramakers in The Netherlands in 1993 as an informal association and production, or manufacturing, entity. They commission designers whose work is as off-beat as their own. The examples here—projects of the "do" group that collaborates with Droog—are expressions of eccentric and often amusing ideas from a collection that premiered at Sapzio La Posteria during the 2000 furniture fair in Milan. Humor also characterized do's promotional literature, in which the images shown here appeared—the sweaty man wielding the sledge hammer and the matronly woman suspended from the lamp are models, not the designers.

facing | "do hit" chair (1000 x 700 x 750mm) by Dutch designer Marijn van der Poll is made from 1.25mm-thick welded sheet steel. The way to assemble it is obvious from its name and a hammer is supplied.

designers say | "Hit, smash, bash to your heart's content, and sculpt for yourself the shape you want it to be. Make it shallow. Make it deep. The choice is up to you. Suddenly, after a few minutes or a few hours of work, you have become the co-designer of the 'do hit...' "

below | "do hit" lamp/swing by Swedish designer Thomas Bernstrand is composed of welded stainless-steel tubing, electrical fittings, and two lamp shades.

designers say | "do attach this lamp... to your ceiling. You are then able to create a different lighting mood to your room by gently letting the lamp swing, or, if you're feeling more energetic, you can make use of the lamp's dual purpose by grabbing onto the handles and letting yourself go. The bigger your room, the more you can swing and the more fun you can have."

date of designs | 2000, produced since
manufacturer | do, Amsterdam, The Netherlands

Photographs by Bianca Pilet.

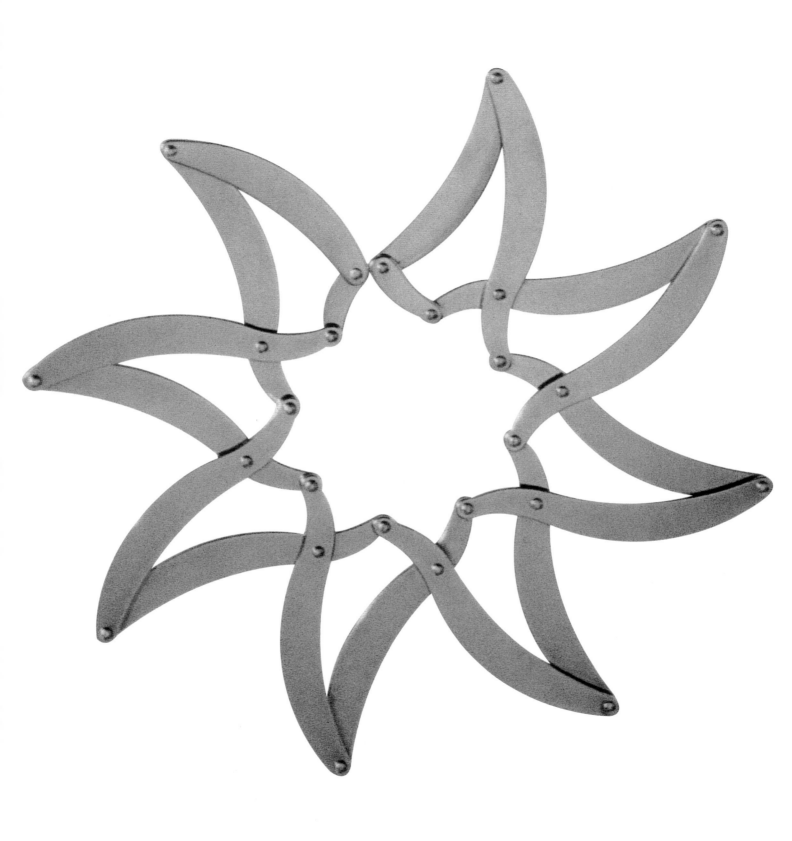

Donato D'Urbino and Paolo Lomazzi
Dinnerware

More than three decades ago, the team of Donato D'Urbino, Gionatan De Pas, and Paolo Lomazzi made design history with its 1967 "Blow" inflatable chair and 1970 "Joe" baseball glove seat. Unfortunately, De Pas died in 1991, but the other two partners, working in Milan, continue to come up with inventive concepts such as the trivet here.

facing | "Augh" trivet (12 x 240mm open) made from 2mm-thick 18/10 stainless steel holds in the heat of a pot or dish placed on it. Formed by cold-pressing machines, the levers are assembled manually. The pivot pins are lathe-turned. All sections are assembled with an orbital riveting machine.

right top | "Augh" trivet completely closed.

right bottom | "Augh" trivet wood prototypes.

date of design | 2000, produced from 2001
manufacturer | Alessi S.p.A., Crusinallo di Omegna (VB), Italy

designers say | "Sometimes, like with this Alessi heat-holder, an idea comes from a completely different direction than the norm. It might start from the end of the process—in reverse gear. For fun or curiosity, we built some quadrilateral wooden models in our studio that were hinged at the vertex, figures capable of being deformed. And we realized that they had extraordinary geometrical qualities that would change greatly when the shape became a parallelogram, whether with identical sides or different sides. And the parallelogram could be formed from two adjacent pairs of the same side. In this last case, we constructed a series of quadrilaterals in a continuous sequence that did not follow a straight line but rather formed a circular star shape! With amazement and astonishment at this extraordinary geometry, we wondered how we could translate it into a real object. Then we realized that it could be made of stainless steel, sheared from a stainless-steel sheet, but that it needed stems to raise it from a table's surface. Next we drew curve after curve until they perfectly joined one other, perfect enough for minimal folding to be possible. Hopefully when people buy the 'Augh' and remove it from the package, they will have the same surprise that Alberto Alessi, the manufacturer, had when we first showed him the prototype."

Photographs by Riccardo Bianchi (right top and facing) and Paolo D'Urbino (right bottom)

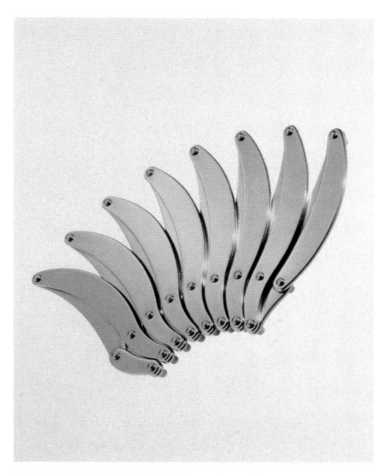

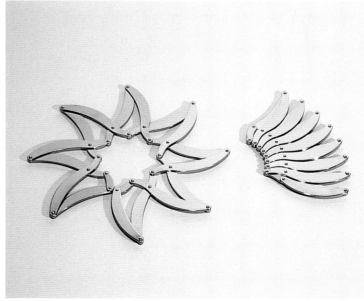

"Double" hollow-ware bowls (200 x 58mm, 250 x 730mm, and 330 x 95mm) are, as the name suggests, double-walled and made from 18/10 stainless steel, with polished surfaces. They are produced with laser-cutting, blanking, cold-pressing, trimming, and surface-finishing machinery.

date of design | 2002, produced since
manufacturer | Alessi S.p.A., Crusinallo di Omegna (VB), Italy

Antti Eklund
Dinnerware

An award-winning Finnish product-and-interior designer, architect, professor, and entrepreneur, Antti Eklund became the founding partner of Animal Design in 1986 to make, market, and sell products of his own designs, such as this stainless-steel tray. His clients also include Alessi and Marimekko.

| "Tracy" tray with "Maximum Brut" pattern (15 x dia. 335mm), featuring incised line art, is made of 18/10 stainless steel. Like a batik, the pattern is first screen-printed on the surface. Subsequently, the object is cold-pressed and then the image is acid-etched. The surface is polished with a rotating silk band. The tray weighs 640g. Jukka Tiainen drew the illustration.

date of design | 1999–2000, produced 2000–01
manufacturer | Animal Design Oy, Helsinki, Finland

designer say | "Our innovation was to combine a traditional stainless-steel tray with the traditional technique of making the small steel signs found on telephone poles and elsewhere. We just used a different pattern on a different scale. The material is glossy, luxurious, and withstands wear and tear. We were also able to create a nice contrast between the hardness of steel and the softness of the image."

Photography by Rami Lappalainen

Maurizio Gamba

Floor tiles

Industrial-grade steel floor tiles can be both attractive and practical in a domestic environment. Designed by a professional—Maurizio Gamba—and available in a number of sizes, these examples are a sandwich of a steel plate and a ceramic. The manufacturer promises that the two materials will not detach over time.

| The floor tiles are available in a wide range of patterns and colors—silvery finishes only are shown here. There are tiles for light-traffic areas (102 x 102 x 8mm or 202 x 202 x 8mm) and for heavy-traffic areas (302 x 302 x 8mm, 602 x 602 x 10mm, or 602 x 1202 x 11mm). Both have grès porcelain bases. The thickness of the AISI 304 stainless steel is 8mm. AISI 316 stainless steel is also available for special uses, for example in hospitals, operating rooms, ventilated walls, and places with high temperatures. The finishes are smooth, decorated, or stiffened. The stainless steel sheet is trimmed with industrial cutters. Presses mold the surface patterns. Automatic assembly machines join the steel to the ceramic with a twin-bond adhesive, mixed in different portions so that the steel and porcelain will remain adhered without detachment.

date of designs | 1999, produced since
manufacturer | Bluestein S.r.l., Pontida (BG), Italy

designer says | "Our steel tiles have only one limitation: the imagination of interior designers. We have been able to create a product for flooring and for any surface that is very easy to lay. While it is suitable for public places, it can also be used in the home—in kitchens, bathrooms, and so on. Some additional features of the tiles are that they are anti-static, bacteria-resistant, and acid-proof."

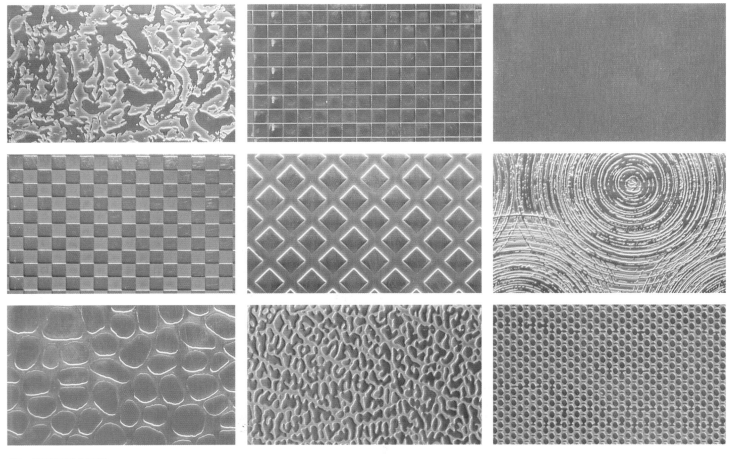

Jérôme Gauthier

Furniture

As with much intelligent contemporary design today, extensive research was conducted by French designer Jérôme Gauthier during his quest for a new concept in case goods. As he explains below, the design is based on the construction of an old-fashioned barrel, with steel banding holding the wooden staves together and allowing expansion.

right | "Band" console (2180 x 720 x 480mm), back view, is built from stretched sheet steel (red epoxy painted) that corsets the amber bamboo body.

below | Front view.

date of design | 2001, produced since
manufacturer | The designer

designer says | "The 'Band' is the result of my graphic and structural research. Balanced between opposites of how it looks and how its works, its conception is revealed in the way that the steel ribbons, traveling around the 'box,' offer both rigidity and flexibility. These lines act as a corset that ties together the volume of the body, just as traditional barrel hoops hold the staves in place."

Stefano Giovannoni
Kitchen- and dinnerware

Manufacturer Alberto Alessi describes the feisty industrial and interior designer and architect Stefano Giovannoni as having a "vague resemblance to a rumpled teddy bear, a man who looks like he might be anything except a much-acclaimed designer." No doubt his reputation has been augmented by Alessi himself who has commissioned Giovannoni to design a large number of products, including those shown here—the most conservative of his œuvre. These products attest to the fine craftsmanship that has continued since 1921 at the Alessi plant in the foothills of Italy's Piedmont region, the area renowned for its metalworking since the mid-18th century.

| "Ethno" baskets vary from dia. 180 to 230mm and from h. 54 to 122mm; the trays from dia. 330 to 400mm. The 18/10 stainless-steel plate is formed by cold-pressing, laser-cutting, blanking, trimming, flanging, and punching machines. The surface is mirror-finish polished. About 35,000 pieces are sold annually.

date of design | 2000–01, produced since
manufacturer | Alessi S.p.A., Crusinallo di Omegna (VB), Italy

---

designer says | "By using ethno figurative shapes, this series revisits and updates the type of piercing that was very common in metal household products of the 1800s. The topologies and the base shapes are the same ones that have previously characterized other families of my products: the 'Girontondo' that played with the icons of childhood memory and the 'Rombi' that had associations with a wicker basket. In the 'Ethno' series, I was interested in verifying—by keeping the same shapes and modifying the piercing—how the image of the products and, consequently, the communication had changed completely."

Photography courtesy Alessi S.p.A.

facing | "Mami" eating utensils comprise a 17-piece set—ranging from a tablespoon to a dessert knife, and including serving pieces (facing top). There is also a compatible kitchen knife set. Over 150,000 pieces are sold annually.

date of design | 2002, produced since

top and bottom | "Mami" pot series is available in 12 sizes with five lids and a four-piece pasta-cooking set (left). The 18/10 stainless steel is formed by blanking, cold-pressing, equalizing, trimming, and welding machines. Surfaces are satin-finished or mirror polished. About 90,000 pots and 2,500 pasta sets are sold annually.

dates of designs | Pot series: 1999, produced since; pasta set: 2001, produced since

manufacturer | Alessi S.p.A., Crusinallo di Omegna (VB), Italy

Photography courtesy Alessi S.p.A.

Ernesto Gismondi
Lighting

Ernesto Gismondi was trained as an aeronautical engineer. He founded the Artemide lighting firm in 1959 and helped Memphis launch its initial presentation in Artemide's exhibition space at the 1981 Milan furniture fair. His firm's 1972 "Tizio" by Richard Sapper may have become the bestselling lamp ever. However, it is the fact that Gismondi also designs award-winning work for his own firm that is noteworthy. The lamp here is available in models for several situations, a common practice of large lighting firms.

facing and below | "Miconos" lamps have been designed for various applications—as floor, ceiling, table (bottom right), and wall (bottom left) models. While the 95mm mouth-blown glass globe may appear to be the main element, the chromium-plated steel hardware (cold-stamped, turned, and tubular) makes the floating illusion possible. A 100w G 95 round bulb is recommended.

right | The chromium-plated steel socket element is fitted through a hole in the globe. The adjusting or hanging handle is molten-applied glass. On other models, such as the sconce (below left), a hole is bored for hanger attachment.

date of design | 1999, produced since
manufacturer | Artemide S.p.A., Pregnana Milanese (MI), Italy

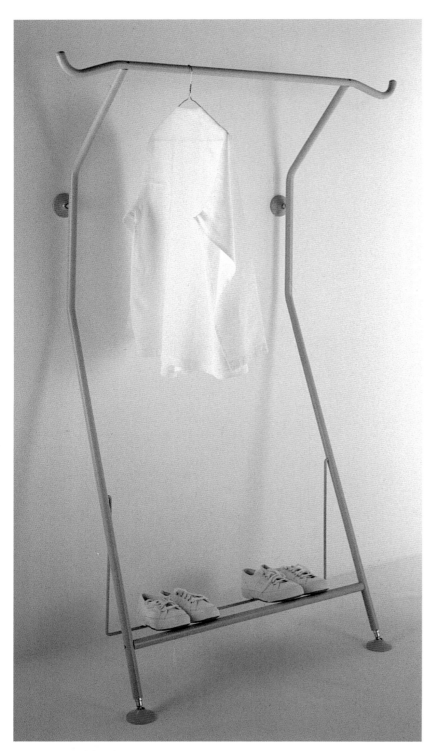

Alfredo Häberli
Furniture

Argentine-Swiss designer Alfredo Häberli, working in Zürich, creates products appropriate to the manufacturers. The clothes rack for an Italian firm employs a somewhat different vocabulary than that of the furniture and furnishings range for a Swiss manufacturer. The successful Trunz Collection (some examples of which are shown here) reveals how variations of form and color in just one material can lend understated vibrancy to a wide range of furnishings that all speak the same language.

facing | "Trunz Collection." Examples here from a wide range of furniture and furnishings offer an insight into the configurations made possible by a single material. The material is 1mm-thick Zinkorblech, a steel sheet electrolytically coated with zinc to prevent corrosion. The fully formed pieces are then powder-coated.

date of design | 2001, produced since
manufacturer | Remo Trunz AG, Arbon, Switzerland

left | "Tauromachia" clothes stand (1250 x 500 x 1700mm), of Driade's Atlantide collection, is configured from bent and gloss-lacquered steel tubes (frame), a bent galvanic steel sheet (shelf), and yellow injection-molded synthetic rubber (feet).

bottom | The completely disassembled stand is made up of only 13 pieces, including the assembly screws, and is easily and economically transportable.

date of design | 1996, produced from 1997
manufacturer | Driade S.p.A., Fossadello di Caorso (PC), Italy

designer says | "The clothes stand uses a minimum of materials for a maximum of wardrobe. The rubber feet are based on those of a cello, just in yellow. Only four screws hold the four tubes together. The packing is very flat (50mm)—in a way, the ideal product for IKEA, but, in this case, for Atlantide/Driade."

Photography by Tom Vack (top) and Betty Fleck (bottom).

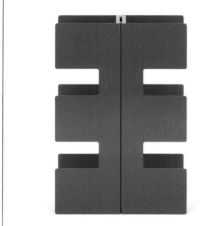

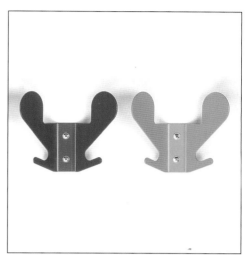

Massimo Iosa Ghini
Furniture

A comic-book-style illustrator early on in his career, Massimo Iosa Ghini of Bologna established a movement in 1983 which he, Pierangelo Caramia, and others called Boldismo. Designs were characterized by biomorphic forms that recalled a kind of 1950s streamline style. While Boldismo never entered common contemporary design parlance, its continuing aesthetic can be seen, if only slightly, in this design by Iosa Ghini for a table base.

facing | "H_2O" table (760 x dia. 1200mm) is composed of chromium-plated or painted tubular mild-steel base. Two mechanical joints connect the pieces of the base that are glued to the glass top.

top | An Iosa Ghini rendering was followed, essentially unchanged, through to final production.

middle | A bending machine forms the steel tubes into precise shapes.

bottom | A close-up view of one of the three points where the base is glued to the top.

date of design | 2000, produced since
manufacturer | Bonaldo S.p.A., Villanova (PD), Italy

manufacturer says | "Like every object in our Biosphere collection, the 'H_2O' table had to express Nature. And steel was the right material to achieve this aim. Meeting both aesthetic and functional requirements, steel is flexible enough to create the sinuous curvatures of the spokes and strong enough to make the table stable in every situation."

Photography by Santi Caleca

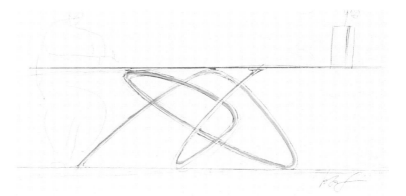

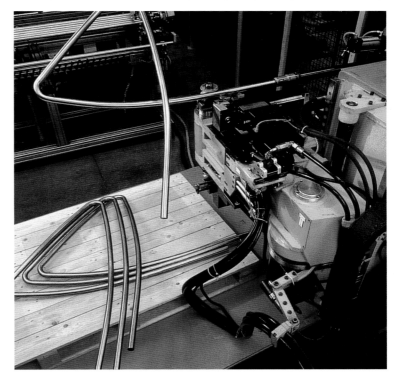

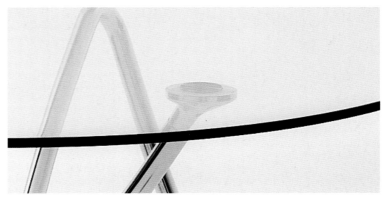

Patrick Jouin

Interior

below right | Chandelier coverings express the designer's fondness for raw materials. The gauze is woven steel organdy. Organdy, or *organza* in French, is a term more commonly used to describe very fine transparent muslin with a stiff finish. The 18th-century-style armchair frames have been metallized by painting with a tinned varnish; the upholstery is an iridescent fabric by Patrick Frey.

below left | The orange-painted metal clock has no numbers. The single rotating indentation indicates the approximate time, possibly best in a place where camaraderie and food are primary.

date of designs | 2000
manufacturer | Chandelier coverings: Delebecque, Lille, France; Clock: Volper-Mellerin, Bagnolet, France

designer says | "The main idea consisted in impoverishing the space, freeing it from luxury, with a view to preserving just the essential features."

Alain Ducasse, the chef and owner of upmarket restaurants in Monte Carlo, New York City, and Paris, opened a second one in Paris in the Plaza Athenée hotel in 2000. For the design, he hired Patrick Jouin, a graduate of the prestigious École Nationale Supérieure de Création Industrielle in Paris and a former employee of Philippe Starck, working on electronic equipment for Thomson multimédia. The Ducasse project challenged Jouin two-fold. He had never before designed an interior and much of the hotel's 18th-century-style dining-room interior had to be retained, particularly the architecture. Jouin had the Louis XV-style armchairs, that could not be replaced, metallized. However, the main new feature in the vast space is the woven metal gauze that sheaths the classical crystal chandeliers. The specially designed undulating felt screen by Poltrona Frau and the imposing metal clock on the mantel are in Jouin's particular language.

Ronen Kadushin
Furniture and accessories

The work here and on the following pages was included in the exhibition "Thinology" curated by Kenny Segal at the Periscope Gallery in Tel Aviv. While producing work for the exhibition, Ronen Kadushin developed a method for forcing a cut-out piece of flat steel sheeting into a three-dimensional form. Thus, he has been able to create objects with the qualities of both two- and three-dimensional objects.

left | "Hole in One" chair (900 x 520 x 890mm), back view. The seating shell (1.5mm-thick steel sheet) is laser-cut with computer-numeric-controlled (CNC) machinery, hand-bent, and spot-welded. The finish is primed and painted. The legs are 16mm-diameter steel tubes. The shocks, made from black synthetic rubber, are connected to the legs with bolts and to the seat with glue. The "Hole in One" is one of a series of chairs.

below | "Hole in One" chair, front view.

date of designs, this page and following | 2000, prototypes 2000 and 2001
manufacturer | The designer

designer says | "Steel is both strong and elastic in thin sheets—the properties I needed to realize my design concept. It is also a familiar material, one to explore to the edge of its potential."

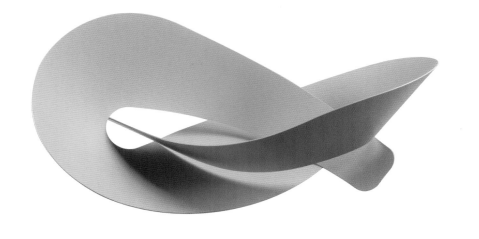

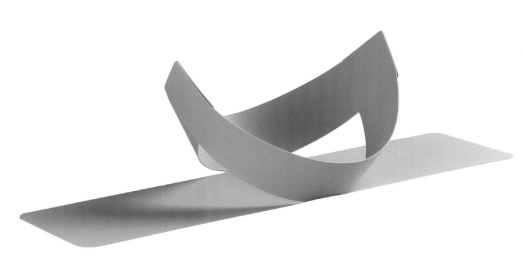

facing top | "High Square Dance" table (600 x 600 x 750mm) is produced using the same methods as the chairs (see previous page). The bright and cheerful colors offset the sturdy material beneath—steel. All Kadushin products are made either in painted 1.5mm-thick plain steel or in 1.2mm-thick mirror-finished stainless steel.

facing bottom | "Low Square Dance" table (900 x 900 x 330mm).

top | "Flat Knot" bowl (250 x 450 x 190mm).

middle | "On Course" bowl (280 x 700 x 180mm).

bottom | "Square Dance" bowl (350 x 350 x 140mm) is an example of a Kadushin design in 1.2mm-thick mirror-finished stainless steel.

Photography by Baruch Natah

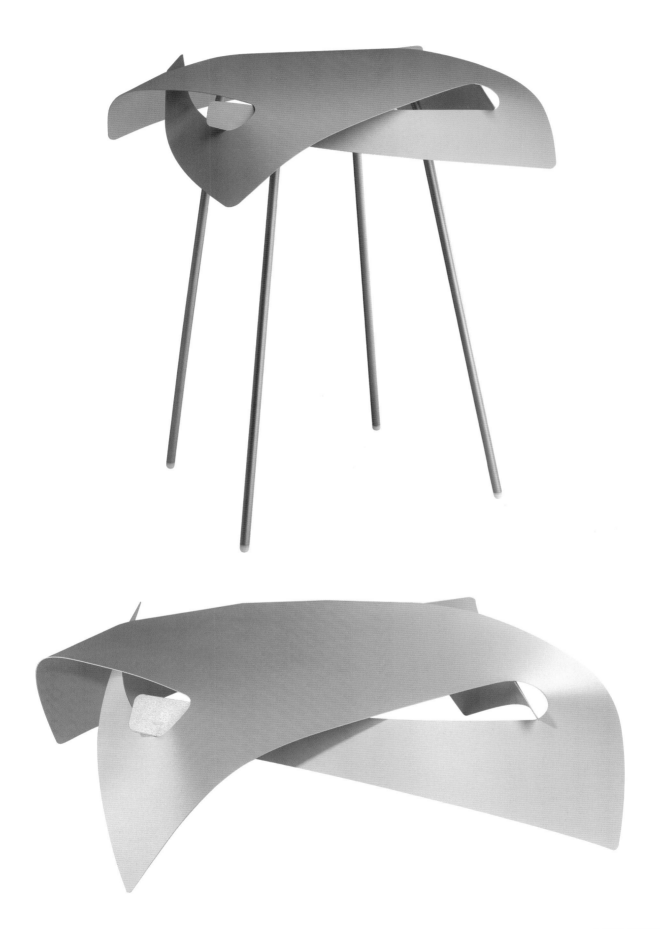

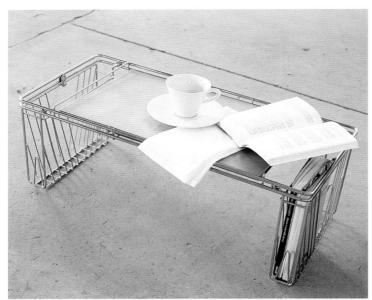

left | "Vassoio Letto" breakfast tray (670 x 305 x 250mm) is made of C40 steel and painted cherry-laminated plywood. The steel rods are drawn, bent, welded, chromium-plated, and screwed together.

bottom left | "Vassoio Letto" book rest (280 x 300mm raised). The hinged surface tilts.

bottom right | "Vassoio Letto" bookshelf (135 x 320–550 x 165mm).

date of designs | 1992, produced from c. 1998
manufacturer | Robots S.p.A., Binasco (MI), Italy

Photography by Aldo Ballo

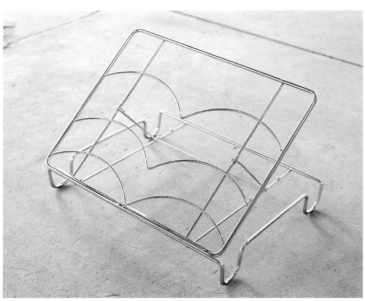

Ritva Koivumäki
Furnishing

The late Finnish designer Ritva Koivumäki was either particularly fond of using steel rods for the production of domestic furnishings or his relationship with Robots fostered the interest. The award-winning manufacturer, established in 1963 near Milan, produces a number of Koivumäki's pieces. Robots's founder, Roberto Rebolini, set up a department in the firm to specialize in metal furniture and furnishings, even designing some himself. In addition to Koivumäki, others have made appreciable contributions to the firm's range, including Bruno Munari with his 1972 "Abitacolo" bed/work/play/living structure in chromed steel rods.

Ralph Krämer
Cutlery and accessories

In order to have use-specific knives for every occasion, one needs a wide array. The examples here by Ralph Krämer are designed to handle the hard-to-soft gamut of the cheese world. Krämer's work, since 1990 for Pott in Germany's renowned forging area, covers a range of designs, from children's eating utensils to a spaghetti fork that will retrieve a single test strand from a boiling pot. All the wares by 100-year-old Pott are hand-made and include, depending on the model, high-tensile, forged blades. Up to 90 different stages of manufacturing—more than 30 for a fork or a spoon—take place.

below | "Marisco" utensils make oyster shucking easy. The knife handle is ergonomically shaped for snug hand-holding. The mail glove both protects the hand and, when washed, will be odor-free, unlike the traditional leather glove. The fork slips under an oyster to keep it intact.

date of designs, pp. 67–68 | 1997–99, produced since
manufacturer | C. Hugo Pott GmbH, Solingen, Germany

manufacturer says | "Today's market is flooded by cheap imports and characterized by declining quality. In contrast to this, a quality product must combine first-class materials, superlative workmanship, and enduring elegance."

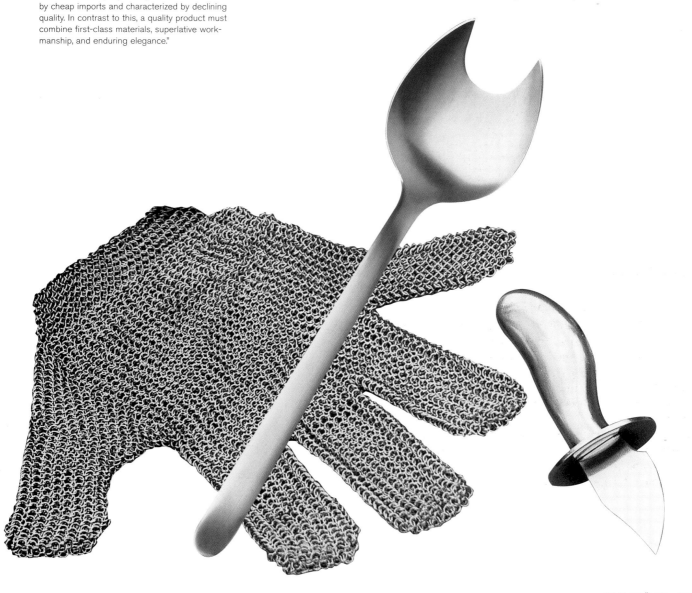

top | "Picado" knife (blade length: 80mm; total length: 160mm) is for slicing parmesan cheese.

second from top | "Frutado" knife (blade length: 55mm; total length: 135mm) is for slicing fruit and vegetables.

third from top | "Raspado" utensil (total length: 240mm) is for slicing semi-hard cheeses, such as emmenthaler, gruyère, and pecarino. It has pointed ends for serving.

bottom | "Manita" utensil (total length: 150mm) separates cheese slices, leaving the server's fingers odor-free.

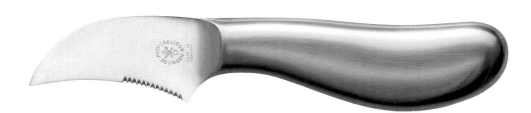

Lemongras Design
Kitchenware

The Lemon partners, Carmen Cheong and Moritz Engelbrecht, studied at the Royal College of Art in London. Carmen majored in industrial design, and Moritz in furniture design. They subsequently established themselves on the Continent, specifically in Munich, because the design climate in the UK was not conducive to finding manufacturers that would produce their work. The example here is produced by their own firm. The pair's concept for a new nutcracker was based on the fact that British and German people normally use a knife to open nuts—not the safest method. Their solution is very simple and, needless to say, economical.

above left | "Gingko" nut opener (without holes, 41 x 41mm) is stamped from a 1mm-thick stainless-steel sheet. The finish is polished.

above right | "Key" nut opener (with holes) is the same size as the "Gingko" and identically produced.

date of design | 1999, produced since
manufacturer | Lemongras Design, Munich, Germany

designers say | "We decided on stainless steel because of its three main properties. 1. It is harder than normal steel and useful when you have a hard nut to crack. 2. After polishing, no other surface treatment, like chroming, is necessary. 3. Stainless steel is the best-quality material for a little tool like this because it gives the product the right image."

Photography by Jürgen Schwope, © Lemongras

Arik Levy
Lighting

Growing up in Israel, Arik Levy studied in Switzerland and subsequently moved to Paris. He works with partner Pippo Lionni. Levy himself designed the fixtures here, as well as others, for French manufacturer Ligne Roset. (For another example of how the filtering nature of steel mesh can be used in lighting, see pp. 156–57.)

facing top row | "Light Pocket" small suspension (320 x 900 or 320 x 3000mm), large suspension (400 x 900mm), and table (750 x 155 x 850mm) lamps include light-filtering diffusers in a stainless-steel mesh which is plasma-welded. The fixture employs a metal of the type employed for industrial filters.

facing bottom row | "Or-la" lamps feature a chromium-plated, stainless-steel chain-link sleeve fitted over the socket and the bulb.

right | "Cloud froissé" bedside (130 x dia. 282mm), table (630 x 155 x 850mm), and reading (175 x dia. 1200mm) lamps include light-filtering diffusers, similar to the "Light Pocket" ones, in woven stainless-steel wire, and are also plasma-welded.

date of designs | 1999, produced from 2000
manufacturer | "Light Pockets" and "Cloud froissé": Ligne Roset, Briord, France; "Or-la": Arik Levy/L design, Paris, France

designer says | "This industrial material [or mesh] permits the passage of light, air, vibrations. I am fascinated by the poetic presence of these mini clouds of intangible light."

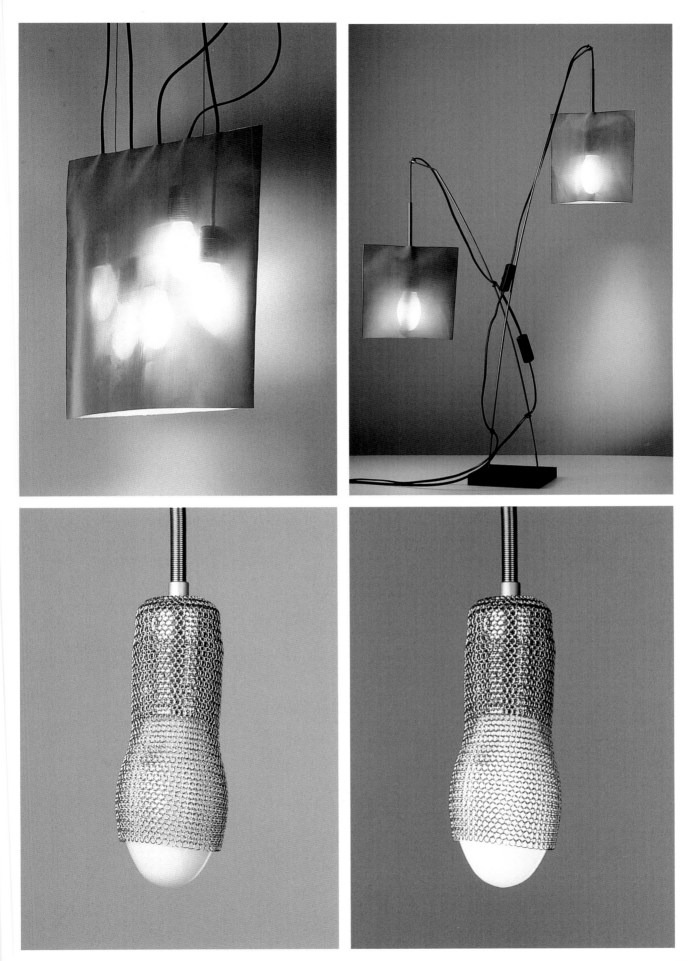

Raviv Lifshitz
Lighting and Furniture

Raviv Lifshitz is an ad-hoc member of the new wave of young Israeli designers who have studied at the Bezalel Academy of Art and Design in Jerusalem. No matter how amusing and seemingly insincere his approach to furniture and lighting may appear, Lifshitz is very serious about his work. Here, every-day, functional objects have been reinterpreted and given a new lease on life.

facing | "Umbrellight" (700 x 700 x 700mm) was first conjured from the brass skeletons of umbrellas but, subsequently, in chromium-plated steel because of the economy and greater flexibility that steel offers. The standard electrical sockets for the four bulbs are connected at the point where an umbrella handle-stick would normally be inserted.

below | A prototype includes the handle-stick that was finally eliminated.

date of design | 2000, produced since
manufacturer | The designer

designer says | "I have an intuitive need to reinvestigate the connections between science, technology, and industry and their reflections in the history of design.... All of my work calls on mass-produced, banal, cheap products [formerly manufactured by others]. The umbrella light, for example, that lost its wrapping remains skeletal, minimal, its own prototype. A new identity has been 'sewn' for it. The final edition of the light is made of steel bars which are more appropriate than brass because, apart from being cheaper, they are stronger and springier."

All photography by Yair Medina

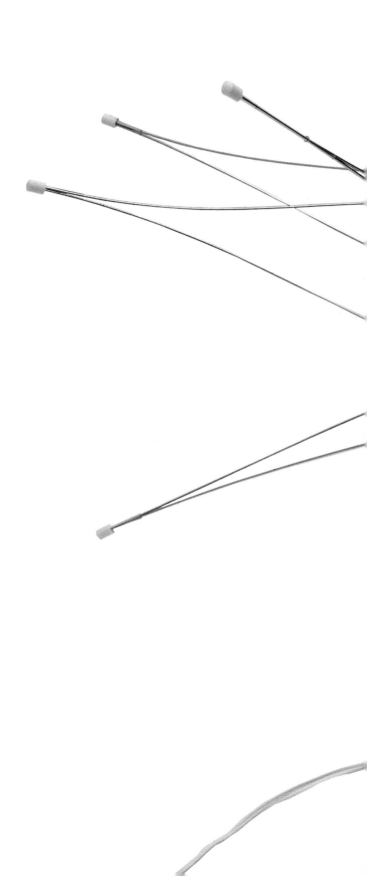

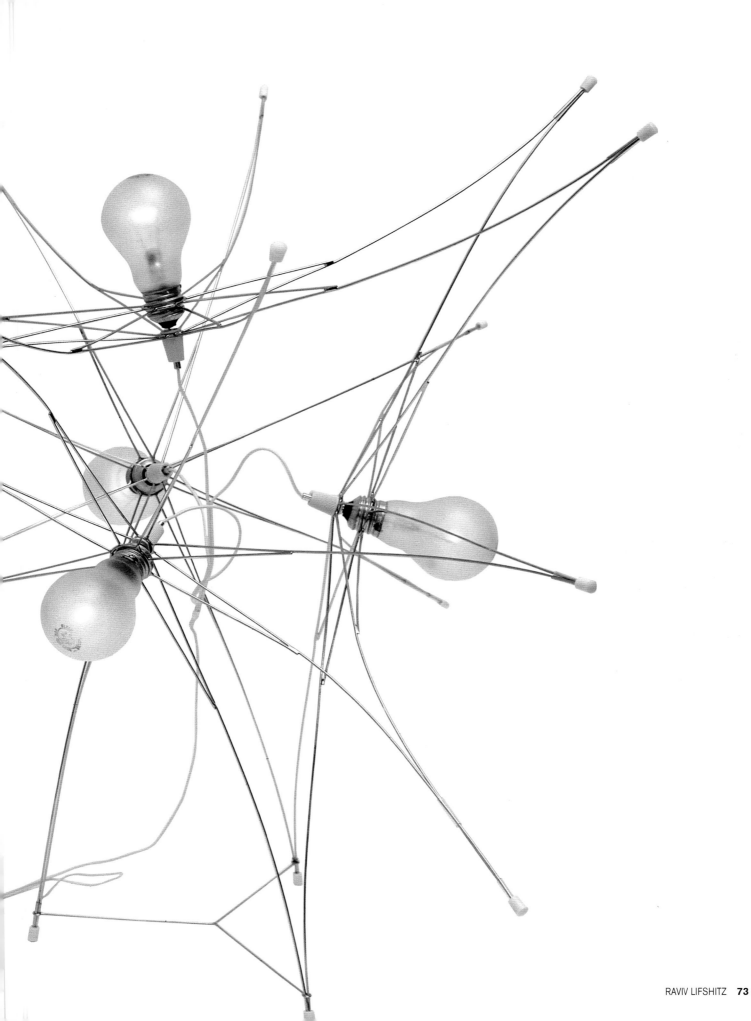

left top | "Ironic" chair (1300 x 800 x 350mm) in molded and high-polished hardened steel (frame), chromed steel (axis), and simulated leather (upholstery). A wing nut into the wood connects the frame axis. The adjusting back device is high-polished molded steel.

left middle row | A close-up view of the chromium-plated articulating leg frame (left). An adjustable-position back bracket on the back permits three different positions (right).

left bottom | The seat and back during production reveal that the construction is more intricate than might be suspected.

date of design | 2000, produced since
manufacturer | The designer

designer says | "I have found that steel ready-made products are unique as materials for another circle of life. The endless character of steel—as a thin constructive bar, a shiny reflector, or a heavy molded element—almost makes it a material with an independent life."

above | "Agency" single-width chair (700 x 900 x 150mm open, 200 x 900 x 150mm folded) is a combination of three ready-made bicycle backracks in stainless steel, designed to collapse. No finish is applied to the metal surface. The springs and spring bars, added by the designer, offer rigidity and flex.

facing | "Agency" double-seat chair combines six bicycle backracks.

date of design | 2000, produced since
manufacturer | The designer

| "Case System" (450, 600, 900, or 1200mm long; 950 or 1740mm high) includes 0.8mm-thick AISI 304 stainless-steel sheets on the doors, façade, and top surfaces. The finish is Scotch Brite™. The case is wood.

date of design | 2001, produced from 2002
manufacturer | Boffi S.p.A., Lentate sul Seveso (MI), Italy

Photography by Duilio Bitetto

Piero Lissoni
Kitchen

A champion of the minimal, Piero Lissoni has designed some of the most refined, pared-down products in the marketplace today. Working in a large architecture, graphics, and product-design studio in Milan, he conceived this kitchen system for a firm that has invested considerable time and funds toward creating sophisticated kitchens, baths, and fittings. To say that they are elegant and expensive may be an understatement.

Michael Lund
Accessories

EyeCatcher is one of a number of firms established by designers to make and market their own ideas – an approach similar to that of some other Scandinavian companies specializing in sleek table-top objects for home and office. In 1998, Michael Lund founded his young firm with the mantra: designs for people whose desire for high aesthetics and simplicity is a way of life.

facing | Wall candle sconce (250 x 250mm) is stainless steel, brushed using two mechanical methods. The candle cup and stem are assembled with glue. The reflection offered by the steel amplifies the candle power. (For another sconce that exploits the reflective nature of steel, see p. 110.)

right | Magazine rack (two sizes: 800 x 150mm and 1500 x 150mm) incorporates inverted "V" spaces cut out by laser machinery for magazine insertion. Bending machinery produces the three folds. Unlike the wall candle's finish, the surface is sandblasted.

date of designs | Wall candle: 1998; Magazine rack: 1999; both produced since
manufacturer | EyeCatcher, Odense, Denmark

---

designer says | "I use steel because I like its purity, its clinical look, and the way you can play with different surfaces, depending on how you finish it. And I like the exceptional way that you can make it look warm by frosting it and cold by making it shiny. Stainless steel is the no. 1 material for expressing simplicity and minimal beauty."

Photography by Michael Lund

Renato Mastella
Staircase

Renato Mastella of Verona designed this rather spectacular staircase that is made possible by an interchangeable and visibly modular configuration placed at the beginning and end of a flight. The modules allow static positioning and can be adapted to various conditions—installed as circular, straight, or mixed versions. Ramp widths range from 360 to 1500mm.

facing | "M16" staircase is circularly configured in this particular installation. The Domex 420 AISI 304/316 stainless-steel module for each step—either galvanized steel or Scotch Brite™ matt-finished stainless steel—is laser-cut and assembled with flat-head, countersunk, or hexagonal screws housed in special disks and convex nuts exclusively developed by the manufacturer. Other fittings include unique turned brushes in iron on one side and special bases to hold various types of steps on the other. Space rings serve to align. The load-bearing structure is made of two lateral self-supporting 16mm-thick trusses, composed of 80/10 steel modular elements. Anchors were developed to attach a flight at top and bottom.

below left | "M16" circular version with clear-glass treads extends from a round aperture.

below middle | "M16" straight version with wooden treads includes a landing.

below right | Close-up view of the modules. The ends of the treads—in this example, wooden—and attachment screws are intentionally exposed at the ends.

date of design | 1995, produced since
manufacturer | Edilco S.r.l, Verona, Italy

All photography by Maurizio Marcato, WALD Studio.

Davide and Attilio Mina
Bathroom fittings

The family team of Mina and Mina has designed a wide range of bathroom fittings that are produced by their eponymous firm. They create a particular design and then apply the aesthetic to a variety of applications, for example a sink, a shower, and a bidet.

| "Toy" is available in a number of models, including one for the shower, sink, or bidet, or as a double-pipe mixer. The sink or bidet models (shown here) include a small handle at bottom left for drain closing. The rotating-lever mechanism adjusts water temperature and force. Made of 100% AISI 316 stainless steel with brushed, polished, or sandblasted finishes, the taps are manufactured with the latest generation of computer-numeric-controlled (CNC) machinery and assembled mechanically with natural gaskets.

date of design | 1999, produced since
manufacturer | Mina S.r.l., Quarona (Vercelli), Italy

designers say | "Through its considerable experience of working with steel, Mina has evolved into a firm that designs and manufactures bathroom and kitchen taps made entirely of stainless steel, the safest and healthiest material available. Thus, the taps are 100% lead- and zinc-free and fully compliant with world eco-regulations which stipulate sanitary fittings to be completely non-toxic in accordance with ANSI/NSF Standard 61 requirements."

Photography by Elisa Negro

Jasper Morrison
Dinnerware/accessories

Designs that are bizarre and unique may garner great attention and provoke delight; more unusual are classic forms with perfect proportions. Jasper Morrison, one of a bevy of British designers who have been sought out by Continental manufacturers, has won numerous awards for his distinctive, understated geometry. The examples shown here would be just as suitable in a living room as in a dining room—the manufacturer suggests the latter.

| "Metal Trend" tray (dia. 350mm), basket (dia. 210mm), salad bowls (dia. 240 and 290mm), fruit/dessert bowl (dia. 120mm), and ashtray (dia. 90mm). 18/10 stainless steel is formed by laser-cutting, blanking, cold-pressing, trimming, and surface-finishing machinery, satinized or polished.

date of design | c. 2001, produced from 2002
manufacturer | Alessi S.p.A., Crusinallo di Omegna (VB), Italy

designer says | "There are few materials which perform as well as stainless steel. It doesn't rust; it doesn't wear out; long-term scratching only makes it more beautiful; it's the material of choice for the kitchen. In London, there are many shops selling equipment for cooks and restaurants, and the best of them are Indian, selling a variety of imported stainless-steel tins, teapots, plates, trays, and so on. That's probably where the inspiration came from to do the Tin Family [for Alessi]."

Stephen Newby
Furniture and furnishings

British designer Stephen Newby may have manipulated stainless steel in a remarkable new manner. Produced through his own seven-year-old Fullblown™ Metals firm, the wide-ranging permutations are dazzling. He has been able to produce squishy pillows with the kind of reflective mirrored surface you might expect from steel but also to imbue them with a rich palette of colors, from reds to deep blues, and create more than 20 textured patterns. Soon he expects to be acid-etching patterns, even images, onto finished forms. Essentially and innovatively, Newby has made steel soft.

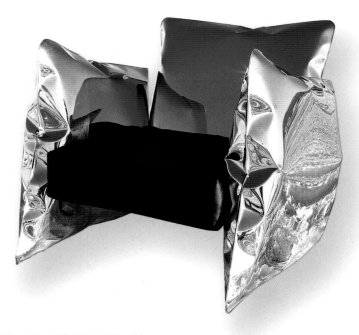

facing and above | "Mosaic Screen" is composed of variously treated stainless-steel skins—different colors and surface textures. The pillows are set within a metal frame. The effect changes according to light conditions, angles of view, and placement. The edge welds of the pillows, or inflatable elements, of all Fullblown Metals products are ground down and blended. The steel skin is quite flexible.

designer says | "The hard, clinical look of stainless steel is transformed into a soft, tactile experience."

top | "Blown Planter" (1100 x 1100mm) in stainless-steel skins with large and small stainless-steel rings is mirror-finished.

bottom right | "Snow Chair" is composed of inflated stainless-steel skin pillows and a leather cushion. The surfaces are mirror-finished.

date of designs | 1996–2001
manufacturer | Fullblown Metals, Ulverston, Cumbria, UK

Donata Paruccini, Fabio Bortolani, and Francesco Argenti
Accessories and furnishings

Italian designer Donata Paruccini created some push pins for Alessi that depart from the dull, traditional ones that are used to attach messages to cork boards. She has also designed some highly functional products with Fabio Bortolani and Francesco Argenti. (See p. 13 for Ron Arad's contribution to the range of virtuallydesign.com, the same manufacturer that produced the combs and bathroom fittings shown here.)

right | "Professional" combs by Donata Paruccini and Fabio Bortolani are chemically cut stainless steel. The models include those that are double-lined (lower left), shaped in the form of a ruler (center left), two-sided to mimic a fish bone (center right), and high-polished to double as a mirror (top right).

date of design | 2001
manufacturer | virtuallydesign.com, Monza, Italy

below | "The Fly" push pins by Donata Paruccini, in stainless steel.

date of design | 2001
manufacturer | Alessi S.p.A., Crusinallo di Omegna (VB), Italy

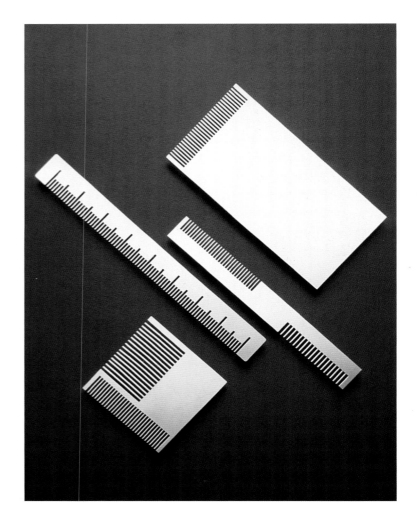

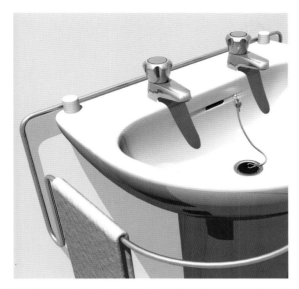

bottom left | "About" towel rail (710 x 520mm) by Donata Paruccini and Francesco Argenti is made of stainless steel. Independently supported by the sink, no wall-mounting or floor placement is necessary. Notice that the maquettes (top left and top right) were hardly altered in the production models.

bottom right | "About" bidet rack (470 x 650mm) by Donata Paruccini and Francesco Argenti is bent stainless steel and, like the towel rack, eliminates the necessity of mounting on a wall or placement on the floor. The design enables convenient towel access.

top left and right | Maquettes.

date of design | 2001
editor | virtuallydesign.com, Monza, Italy

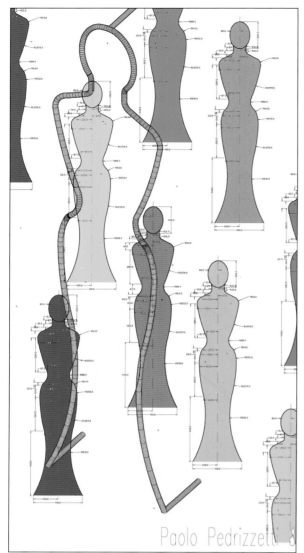

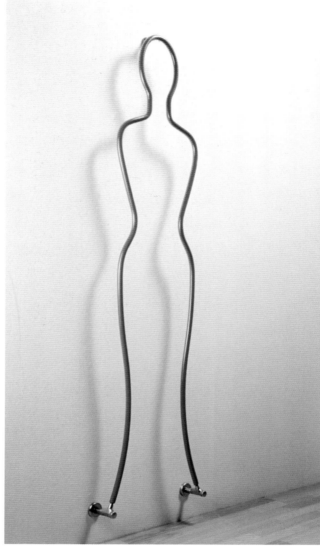

Paolo Pedrizzetti & Associati
Clothes warmer

Bent steel tubing and a source of heat make for imaginative and humorous possibilities—here, a robe warmer in the silhouette of a woman by a Milanese design studio. Even its name, "MissHot," is amusing. Electricity or warm water heats up the one-piece pipe that can then warm a bathrobe hung on it.

designers say | "To furnish, to dry, to decorate, to play, to warm, to dream. How many things can be done with a colored 3.5m-long stainless-steel tube and a drop of water or a little electricity? Cold, hard stainless steel has been gently shaped as a female body, heated by warm water or by 80w of electricity. After a tonic shower or a relaxing bath, there is nothing better than having a warm bathrobe at hand. And what might be better than a 'body heater'?"

above right | "MissHot" radiator (400 x 1200 x dia. 20mm) is a shaped stainless-steel tube, 3500mm long. Bending and computer-numeric-controlled (CNC) machinery are used. The finish is powdered polyester. The system operates by 80w of electricity or by hot water.

above left | A drawing gives an insight into the amount of attention paid to the production parameters.

date of design | 1998, produced from 2000
manufacturer | Tubes Radiatori S.r.l., Resana (TV), Italy

Photography by Studio Fuoco

Permafrost

Cabinet lock

Despite the fact that this object has only been realized as three prototypes, the idea is a clever and appealing solution to a vexing problem: theft. The design was developed by young designers Tore Vinje Brustad and Eivind Halseth who, with Oskar and Andreas Murray, formed the Permafrost studio in 2000 in Oslo. The simple mechanism is intended to prevent the unauthorized opening of shop cabinets containing valuable goods.

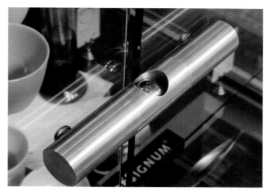

top to bottom in sequence | The cabinet handle (150 x 100 x dia. 25mm each side) is composed of two bars and a ball, all in stainless steel. The ball is 20mm in diameter. Soft rubber gaskets separate the steel from the glass cabinet door. An acrylic "key" (shown in the second image) is used to push up the ball, making it possible for the two sides of the bar to be separated.

date of design | 1998
manufacturer | The designers

designers say | "Since the steel ball is the vital part of the construction, stainless steel was the natural choice for the other parts. The fact that stainless steel gave the design a feeling of quality, durability, and precision didn't do any harm either."

Photography by Tore Vinje Brustad

Roberto Pezzetta/Zanussi Industrial Design
Washing machine

This washing machine was designed by Roberto Pezzetta and the in-house design staff of the almost-90-year-old Italian manufacturer Zanussi, which was incorporated into Electrolux in 1984. The "Jetzi" all-steel machine features a proprietary method of clothes cleaning. The Jetsystem, as the name suggests, "spray washes" the garments and the advanced, fuzzy-logic controls automatically adjust the water level, electricity consumption, wash-agitation force, and length of time required to deliver optimal wash results, irrespective of the fabric and soilage. The pitched front makes access to the inside easier. For its products, Zanussi has received a number of prestigious awards.

facing | "Jetzi" washing machine (596 x 850 x 620mm) is shown in x-ray-like detail.

below and right | Incorporating the manufacturer's ZF 1110 JX Jetsystem, the parts of the "Jetzi" are DC 04 sheet steel. The steel cabinet is pickled and powder-varnished. The structure is screwed together.

date of design | 1999, produced from 2000
manufacturer | Electrolux Zanussi S.p.A., Porcia (PN), Italy

manufacturer says | "From the beginning, we have been using steel for washing machine cabinets. It's the most suitable material for this item, as far as mechanical and economical considerations are concerned."

Piano Design Workshop

Kitchen fixture

Might a sleeker, more sensuous design for a kitchen burner unit exist? Maybe. Maybe not. But certainly this one-piece sheet-steel interpretation by the Piano Design Workshop in Genoa proves that steel can be an aesthetically noble material.

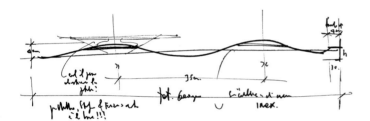

facing | "P705–5" burner (720 x 510mm), shown here in close-up without the risers, is made from 18/10 stainless steel (AISI 304 BA), including the control knobs. High-powered machinery was used to press a single metal sheet into shape.

left | A preliminary sketch of the "P705–5" burner.

below | The unit is mechanically fixed without visible connectors or screws.

date of design | 1997, produced from 1998
manufacturer | Smeg S.p.A., Guastalla (Reggio Emilia), Italy

designers say | "The material of the "P705–5" is very important. The hob is made entirely of steel which creates the effect of great unity, luminosity, and elegance and guarantees solidity and a long life. A clean, basic object decorated by the light it reflects, the hob clearly expresses its essential qualities: simplicity, durability, cleanliness, and beauty."

Photography copyright © Smeg S.p.A.

Bruna Rapisarda
Bathroom fixture

Italian designer Bruna Rapisarda, working from her studio in Milan, created a bathroom sink-and-counter combination that, as its name "Box" suggests, is an inverted stainless-steel container. Surprisingly, the sink, top, and sides are a single piece of unwelded metal.

| "Box" (650 x 450mm or 450 x 450mm) washbasin and countertop is made from one sheet of 18/10 stainless steel, bent and pressed. The tube for the towel is screwed to the ends.

date of design | 2001, produced since
manufacturer | Regia S.r.l., Taccona di Muggio (MI), Italy

---

designer says | "Stainless steel is a strong but light material, and it represents the qualities of hi-tech and minimalism in design."

Photography by Enrico Suá Ummarino

Rapsel
Bathroom fittings

Since its founding in the 1970s, Rapsel has never apologized for catering to "a sophisticated public consisting of opinion leaders and consumers who lead a stylish lifestyle and have good taste." Their stable of freelance designers has included Philippe Starck and the late Shiro Kuramata, in addition to those whose work is shown here. Rapsel also makes showers, mirrors, and cabinets.

facing | "Lake Carezza" rectangular tub (940 x 1850 x 515mm) by Peter Büchele is double-walled AISI 304 stainless steel (3mm thick). The outside is satin-finished; the inside polished. Also available with both sides either fully satin-finished or fully polished.

date of design | 2001, produced from 2002

top and middle | "Wing" hand basins (330 x 100mm and 660 x 100mm) by Gianluigi Landoni are made as double, corner, or with-extension models. AISI 304 stainless steel is satin-finished outside and polished inside.

date of design | 1996, produced since

bottom | "X-treme" sink (450 x 315 x 80 or 750 x 485 x 100mm) by Peter Büchele, with a built-in syphon and waste fitting, is polished AISI 304 stainless steel. ("Miss Fer" mirror, 1999, designed by A. Frezza and F. Roselli.)

date of design | 1998, produced from 1999

manufacturer | Rapsel S.p.A., Settimo Milanese (MI), Italy

designer says | "Steel is a material that has always fascinated me with its expressive potentiality. Its technology also fascinates me—from the extraction of the metals that compose it to the last phases of working with it. Steel has never been fully investigated for its expressive capabilities. Hence, using it for a wash basin continued my fascination…. On paper, we made drawings [for the "Wing" basin] that would eventually result in the shaping of a sheet of steel that was like flowing water." —Gianluigi Landoni

designer says | "I like straight shapes, cubes, cylinders, and cones that come about by simply shaping sheet metal. High-grade steel is an homogenous material, one whose surface allows a high mirror finish, creating the sensation of lightness. Products in it seem almost transparent. And high-grade steel is extremely resistant to the effects of time. Marks of usage represent vitality rather than damage. That's why I create high-grade steel objects." —Peter Büchele

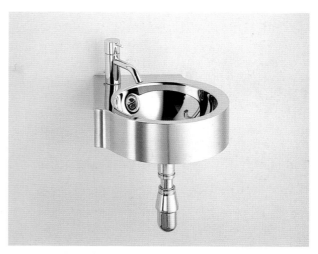

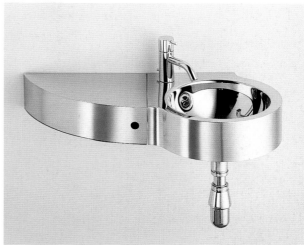

Karim Rashid
Furniture and furnishings

Born in Cairo, Karim Rashid moved with his siblings and English/Egyptian parents to Canada where he subsequently studied design. He later emigrated to New York and established his own studio in 1993. Rashid now claims a large roster of clients and published the 2001 volume of his work, *I Want to Change the World*. The chairs and coffee table shown here, like much of his work, have been infused with a vivid sense of color that begs for the wearing of sunglasses and reveals a predilection for sensuous forms.

facing | "FloGlo 1" and "FloGlo 2" chairs and "Lexicon" coffee table are made of 14-gauge (2mm) bent and welded steel. The surface is painted with high-gloss lacquer, white outside and florescent inside. The tubular legs and articulated glides are chromium-plated steel.

right | "Wysk" coat stand (1800 x 500mm) comprises eight identical sections, made of bent 15mm-diameter chromium-plated steel tubing.

date of designs | 1997
manufacturer | Idée Co., Ltd., Tokyo, Japan

designer says | "Steel is a material that you can't challenge. I can't be disputatious with steel—an ubiquitous, gargantuan resource. I love steel. I have made so many frames, chairs, chaises, stools in steel. It is easy to use, easy to fabricate, economical, and structural; it has low tooling costs and the highest material-and-strength ratio [compared to some other metals]; it is practical and frugal; and it can be finished in a multitude of ways."

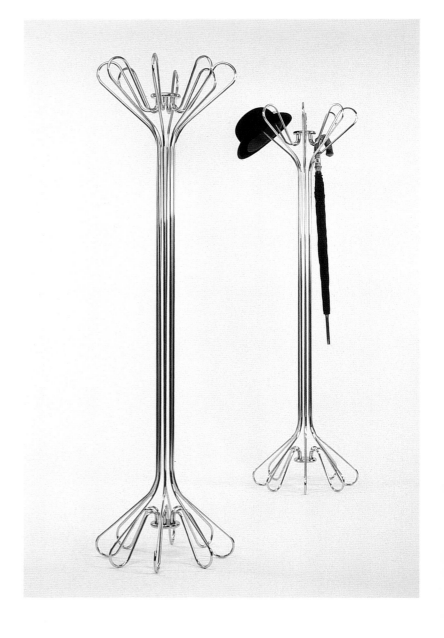

facing | "Mekano" cookware (casseroles: dia. 20, 30, 30, and 50mm; saucepans: dia.12 and 20mm; frying pan: dia. 24mm) are made with 7-ply material: 18/10 stainless steel for the top layers and another metal for the core. The bowls (dia. 20, 25, and 40mm) are all 18/10 stainless steel. Deep-drawing, welding, polishing, and grinding machines; lathes; and eccentric pressing machinery are used for the construction. The handles are welded on.

below | "Halo Saunatonttu 4" (1090 x 500 x 500mm; 50kg) includes a housing in Aluzink. (Aluzink, which resists corrosion, is a proprietary cold-rolled steel with a zinc/aluminum coating; the zinc content offers resistance to scratches.) The stove will efficiently heat an 11 cubic meter area, has a low surface temperature, and is safe if children are around. When opened, the lid can be fully vertical. The compartment holds 100kg of rocks, thus creating a humid but gentle heat.

dates of designs | "Mekano" cookware: 1999, produced since; "Halo Saunatonttu 4": 2000, produced from 2001
manufacturers | "Mekano" cookware: Opa Oy, Mikkeli, Finland; "Halo Saunatonttu 4": Saunatec Group, Hanko, Finland

---

designer says about the cookware | "There are two things I hate about pots and pans. There is never a proper place to put a lid, and, when storing pots and pans, they always end up on a bottom shelf, and you find yourself crawling on the floor to find them. Therefore, I designed a wall-mount system that stores the pots and pans."

Photography by Ilmari Kostiainen (cookware) and Jukka-Pekka Asikainen (sauna stove)

Ristomatti Ratia
Cookware and stove

"Risto" Ratia, the son of the founder of Marimekko, has designed a range of attractive pots and bowls, with all elements in stainless steel, that is part of a wall-hanging system. He also designed the "Saunatonttu," which in Finnish means "Elf sauna." Today there are about 1.5 million saunas for Finland's population of 5 million, a testament to the popularity of hot bathing that began at least a thousand years ago. This model is technically advanced and easy to use.

7-ply material, top
layers 18/10 stainless
steel, core mixture of
aluminium for optimum
heating. Suitable for all
hotplates and ovens.
Copyright 1999
Ristomatti Ratia/OPA

Alon Razgour
Desk accessories

Similar to the desktop accessories by the Blu Dot group (see pp. 24–25), Alon Razgour's designs use unpainted stainless steel and have a variety of perforation characteristics. Razgour wants the user—who purchases the products completely flat—to participate in the "design" of the stands, fruit bowls, utensil racks, letter boxes, and other such utilitarian objects. An Israeli, Razgour has also designed electronic products, such as communication devices and computer peripherals, for clients including Ericsson, Charlotte's Web Network, Rooster, Siemens, Sital, and Tadiran.

facing and right | "Two Dimensions to 3D" (various sizes and configurations, flat and folded) is a stainless-steel sheet.

date of design | 2000
manufacturer | The designer

designer says | "Every plate exists independently as a two-dimensional element and turns into a functional product by folding from two dimensions to three. Every product can be folded into many different variants, a feature that interests me. All the facets of disassembling and assembling as well as my inability to control the final product completely and my lack of knowledge about the end product fascinate me. However, users cannot alone decide where to posit the product—they and I are partners in creating the results. (And I ask these users to use their imaginations.) The thought that I might in the future meet, come across, or see a continuation of my process fascinates me and feeds my inquisitive nature."

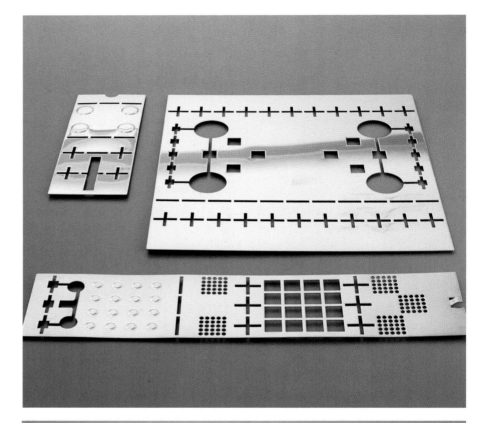

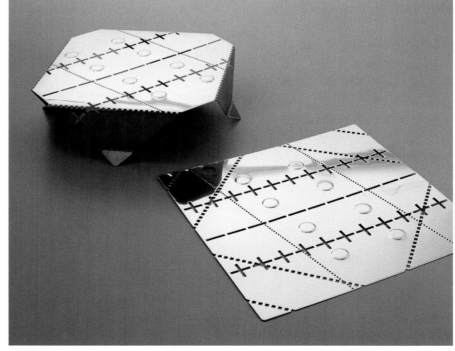

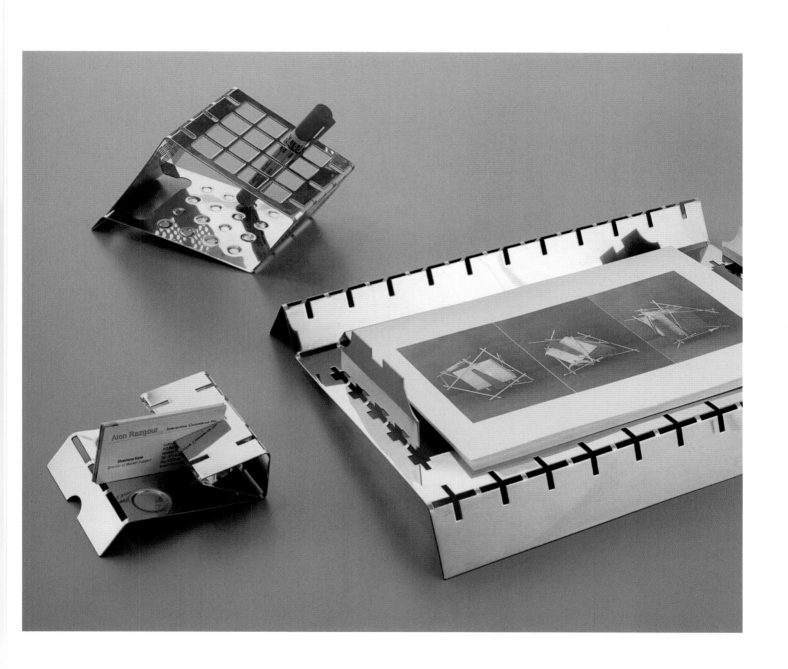

Ely Rozenberg

Furniture and kitchenware

An Israeli-Italian designer living in Rome, Ely Rozenberg, along with Italian professor Vanni Pasca, has done much to further recognition of contemporary design in his native land through the travelling 2000–01 exhibition "Industrious Designers: Young Israeli Designers." His own work shows a fondness for the zipper and rejects the normal welding, riveting, and soldering techniques for making steel furniture.

facing and top row below | "Moby" chaise longue (1400 x 600 x 600mm) is made from 0.6mm-thick laser-cut harmonic (flexible) steel sheets, a type of steel that is used to make springs. The fabric is sewn to an adhesive plastic tape and the zipper. The tape is attached to the steel with an epoxy glue.

bottom | "Tamnun" centerpiece (585 x dia. 70mm) is made of harmonic steel (like the chaise) and has eight zippers.

date of designs | 2000, produced from 2001
manufacturer | The designer

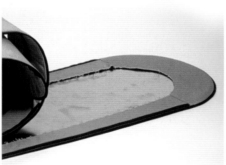

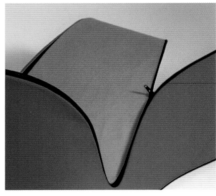

• | A zipper encircles the periphery to close the sections and form the chaise.

designer says | "Why flexible steel for the zipping system? Because it behaves like a plastic material but is far more resistant and long-lasting. And the use of steel is based on the idea of assembling furniture in a quick way, connecting pieces of thin, flexible (0.6mm) steel with a zipper. It is possible to create a type of nomadic furniture that mediates between the need for comfort and the small space available in flats [apartments], and it follows the continuous possibility of moving from one home to another that characterizes the third-millennium lifestyle. With this system it is possible to make any type of furniture—sofas, chaise longues, etc.—without a skeleton. The volume is created when the flat sheets of steel are assembled by simply closing the zipper. The furniture is easily folded, moved, and reassembled with the zipper. The steel can be covered with any kind of material."

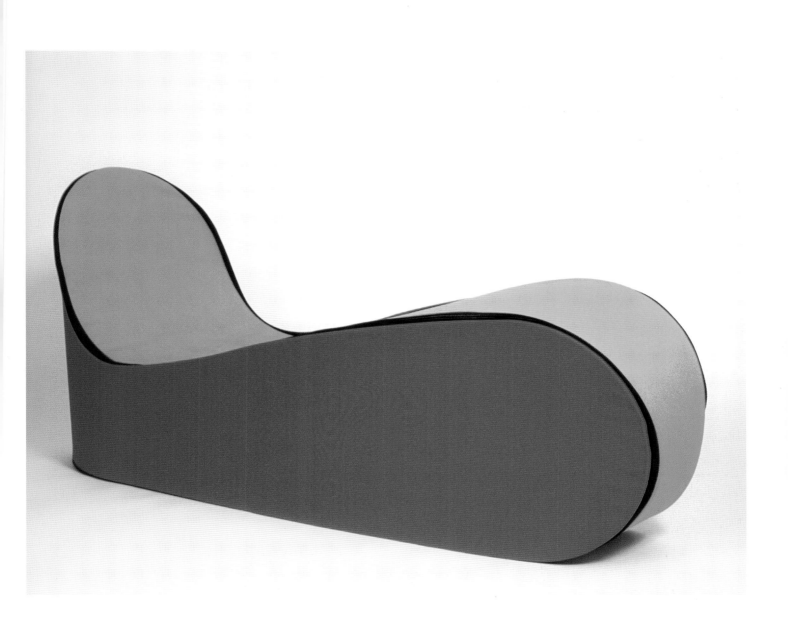

Runtal
Radiators

How many ways can a radiator be designed? Judging by the examples here, the possibilities are infinite, with forms taking on additional roles as artwork, dressing mirror, unadorned column, and decoration in a young person's room. Those shown here are only four of more than two dozen models available. Runtal, the inventor of the panel radiator in the 1950s, also offers versions in the form of stair banisters and invisible baseboards.

facing | "Striane" (max. 2200 x 9550mm) is a steel-welded construction, available in 80 colors, polished or satin-finished.

right top | "Flip" (980 x 1980mm; 754w) in the Ferrum Collection includes a radiator element that is steel and cast iron and RAL colored.

bottom left | "Fuego" column (1700 to 2500 x dia. 220mm; 508 to 748w) in stainless steel is possibly the manufacturer's most simple design. It convects air through piercing in the base.

bottom right | "Versus Mirror" (1736 x 478mm), for a bathroom or bedroom, includes a mirror, light, and shelf.

manufacturer | Runtal Italia S.r.l. (Zehnder Gruppo), Lallio (BG), Italy

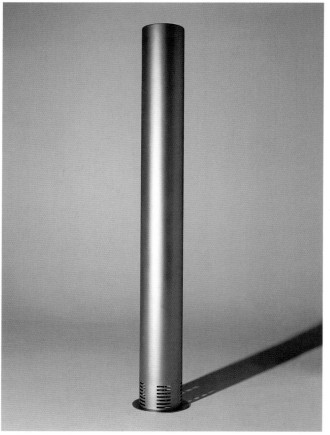

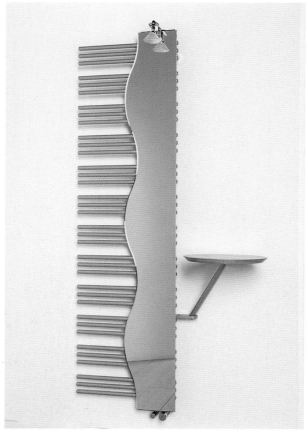

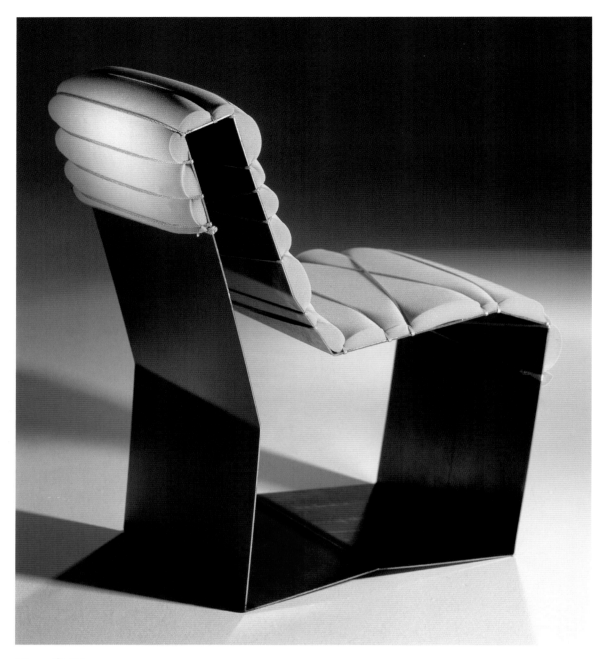

Timo Salli
Furniture

Timo Salli was an original member of the four-person Snowcrash group of Helsinki that first showed its work in 1996 and whose unusual ideas, such as an inflating floor lamp, soon attracted wide press attention. The group's business efforts were taken over by the firm that also owns Artek, the producer of Alvar Aalto furniture. The holding company also incorporated Valvomo, and some of the Snowcrash members now work for the firm. However, Timo Salli has been active in his own studio since 1993 as a product designer, architect, and manufacturer of his own products.

facing | "Power Ranger" chair (prototype) is sheet steel and unfinished polyurethane attached with cording.

above | "Zik Zak" folding chair (prototype) is composed of steel structural members. The upholstery is an acrylic-plastic-infused stainless-steel fabric. The chair was designed for a 1997 Snowcrash exhibition.

date of designs | "Power Ranger": 1996; "Zik Zak": 1997
manufacturer | The designer

designer says about "Power Ranger" | "The flexibility of steel is the motor that makes this chair rock."

designer says about "Zik Zak" | "Collapsible chairs are usually very fragile and lightweight. However, the 'Zik Zak' is strongly constructed and features a playful mechanical device. A collapsible chair need not be less; by function alone, it can be more. How the chair operates is uncamouflaged; there is a clearly visible, baroque mechanism."

Simplicitas

Accessories

The letter scale on the facing page was the first product in the Simplicitas collection, which today includes the work of a number of different Scandinavian designers. The scale, named "Epistola"—Latin for "letter"—proved highly successful, with sales of more than 100,000 pieces.

designer says | "The choice of stainless steel was made because of its high quality as well as its aesthetic."
—Thomas Dahlgren (Simplicitas)

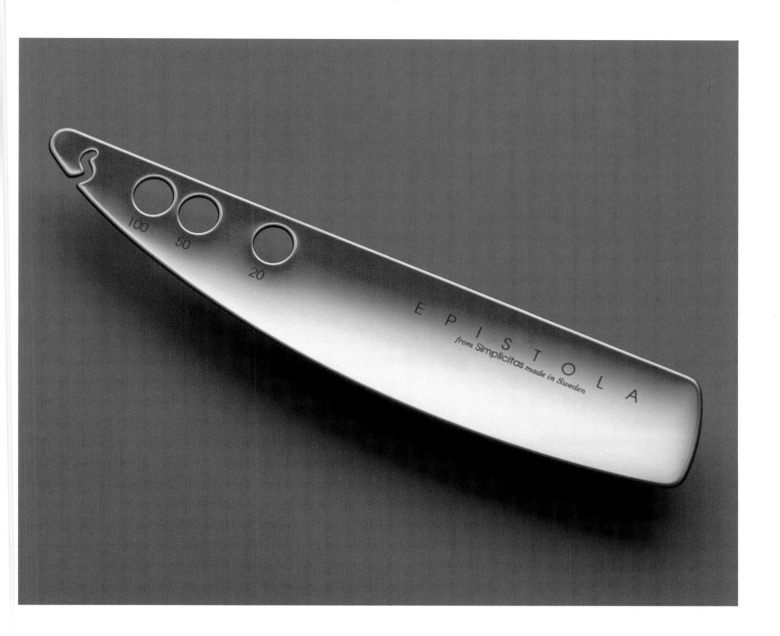

facing top | "Candela" (322 x 95 x 49mm)
by Louise Enlund is stamped from a stainless-
steel sheet, as are all the products here.

facing left | "Veritas" mirror (95 x 62 x
0.8mm) by Eva Åbinger is, of course, highly
reflective and unbreakable, and therefore
particularly suitable for children.

facing right | "Lamina" by Olof Söderholm is
for slicing cheese and butter.

above | "Epistola" letter scale (149 x 29 x
1mm) by Teo Enlund.

date of designs | "Candela": 1996, pro-
duced since; "Veritas": 1997, produced from
1998; "Lamina": 1995, produced from 1997;
"Epistola": 1994, produced from 1995
manufacturer | Simplicitas AB, Stockholm,
Sweden

Photography by Joakim Bergström.

Chris Smith and Susan Kollmer

Photo/art frame

Intended primarily for photos, this frame can display art or any other image. So simple is the concept that little explanation of how it works is required. (See pp. 122–25 for another Smith product.)

designer says | "We could have used a plastic, but stainless steel offered a more refined finish." —Chris Smith

facing | "piclip™" frame is a single sheet of stainless steel, punched out and brushed.

below left | The springy nature of a thin stainless-steel sheet allows the fold-down part of the one-piece unit to serve both as a stand and a lever to keep a photo or other image in place.

below right | In the stamping process, a "U" shape is punched out in one direction, and, within that, another "U" shape is punched out in the opposite direction.

date of design | 2000, produced from 2002
manufacturer | Piclip Ltd, London, UK

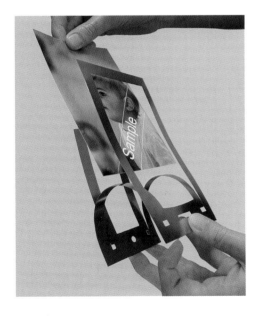

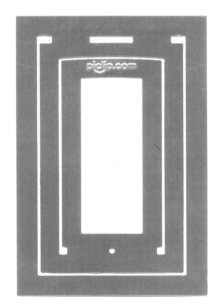

Henk Stallinga
Furniture, accessories, and trophies

Henk Stallinga's work (see similar examples on pp. 42–43 and 148–153) provides an insight into the kind of thinking and ideas of cutting-edge designers in The Netherlands today. Offbeat to say the least, these examples are taken from larger bodies of work. Although some of his pieces will raise a smile, Stallinga is nevertheless a serious designer.

facing | "Set Seat" (800 x 2200 x1500mm), shown here at a 1993 exhibition of the designer's work at the Gerrit Rietveld Pavilion "Interior," is 5mm-thick industrial steel sheet, bent four times to form a bench. The extending "carpet" section acts as a counterweight.

below | "Computer Terminal" (1000 x 1000 x 500mm) is 8mm-thick industrial steel sheet that is enamel-painted and sprayed with a transparent coating. The structure has a "give."

date of designs | "Set Seat": 1993, produced since; "Computer Terminal": 1994, produced 1994–96
manufacturer, pp. 114–17 | Stallinga BV, Amsterdam, The Netherlands

designer says | "These computer terminals are appearing in public places in Amsterdam like libraries and airports. Steel is vandal-proof."

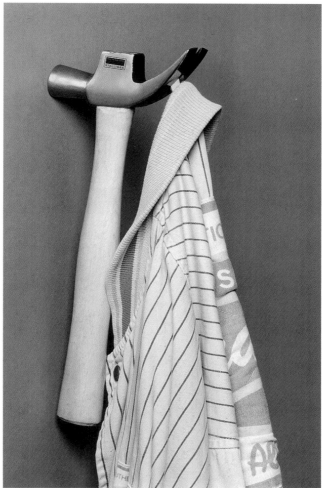

facing | "European Engineering Award" (200 x dia. 100mm) includes a stainless-steel base and a length of wire bent into the silhouette of a traditional trophy. When the motor in the base spins the wire around, the illusion of a three-dimensional object appears. (220v motor, 100 revolutions/min.)

above | "Coathammer Jut" (330 x 130 x 35mm) marries a regular hammer, the head of which is made of tool steel—the hardest available—with a nail. A special drilling machine bores the hole for the nail that is then clamped into the head of the hammer. When complete, the user forcefully smashes the hammer, with the nail extended, into a wall. The claws of the head then act as hooks onto which clothes can be hung.

left | "Clojo" toilet-paper holder (220 x 120 x 3mm) is a reformed standard steel wire coat hanger that is galvanized and polished. The form is made with a setting machine.

date of designs | "European Engineering Award": 1998, produced since (every two years; 15 pieces). "Coathammer Jut": 1997, produced 1997–2000. "Clojo": 1999, produced since

designer says about "Coathammer Jut" | "For the construction, you need the worst possible steel hammers in order to be able to drill holes into the steel. The difficult part is finding these very bad-quality hammers."

Photography by Renee van der Hulst ("Set Seat" and "Computer Terminal"), Hans Petersen ("Coathammer Jut"), and Rene Gerritsen ("Clojo" and "European Engineering Award").

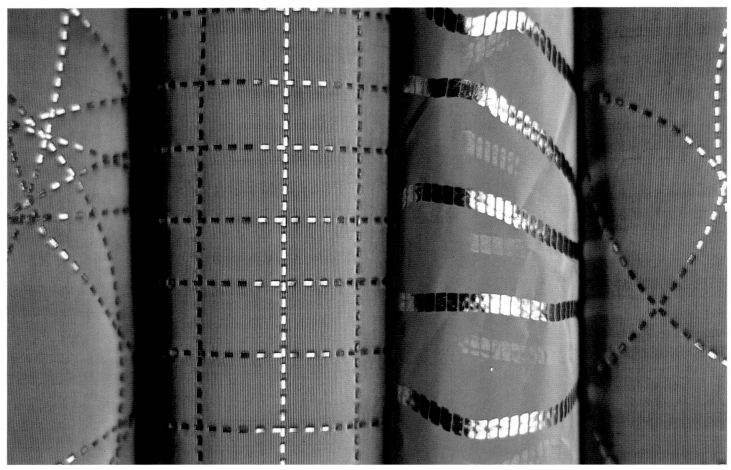

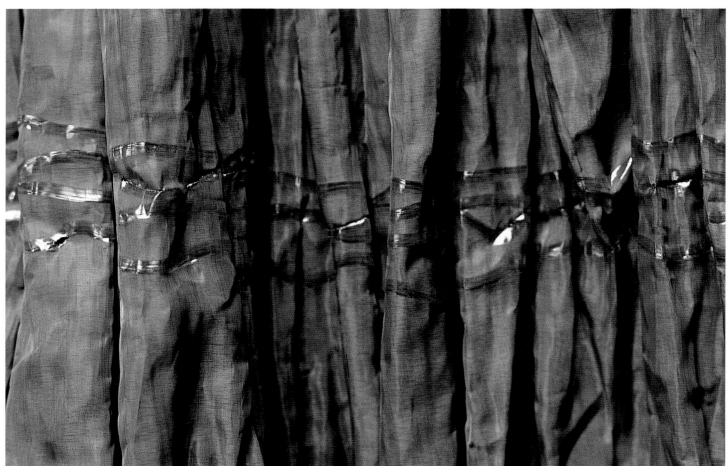

Janet Stoyel

Textiles

British artist Janet Stoyel was already pursuing her imaginative work with metal fabrics before she attended London's Royal College of Art. Continuing her research in her own facilities in Devon, she has been able to create unique and quite remarkable effects on steel weavings through the use of self-developed, sophisticated electronic equipment and laser-cutting machinery. The process, which is a difficult one, involves the material being woven to precise specifications by an anonymous firm in Asia.

designer says | "Metal has an intrinsic beauty unsurpassed by any other material. Stainless steel, in particular, has a timeless quality which can be exploited to the full with each new generation of aficionados who suddenly discover its almost sensual appeal. I use stainless steel partially for aesthetic reasons but mainly for its latent design possibilities. My ultrasonic process brings the hidden elements of the material to bright, shining life. And my stainless steel is completely fireproof, does not tarnish, withstands weathering, looks fantastic, and is available only from me!"

above | Patented ultrasonic equipment produces an infinite range of patterns—with flat or crinkled surfaces—on an AISI 304 or 316 stainless-steel mesh. Vibratory techniques applied at a high resolution act as an "alchemistic" catalyst and create brightly colored interference-spectrum patterns that flash or trail across the surface of the mesh. When the sound waves pass through the material, its molecular structure is radically altered. No dyes, chemicals, or wet processes are used in the creation of the color patina. Even so, environmental ethics are respected in all production stages. The piercing is laser-cut. Pieces can range from 3000 to 30,000mm in the soft mesh and from 350 x 350mm for heavier weights. (The color range here is but one of many possibilities.)

facing | A hand-controlled lower-vibration process produces bright, burnished trailing patterns that appear to be stamped but are not. The drawn quality of the lines is permanent and can be suited to client requests.

date of designs | From 1994, continuing and according to client specifications
manufacturer | The Cloth Clinic, Shelton, Devon, UK

below | "YLS 115G Jet Black" watch
(33 x 38.05 x 9.7mm) has a band and
case in 316 L stainless steel produced
by the MIM (metal injection molding)
process. Generally, the MIM technique,
developed in the 1970s in the US,
involves a metal powder, mixed with
organic binders, that is injected into a
mold. The process, which can form com-
plex, intricate parts in large quantities and
offers a fine surface finish, produces a
metal with the characteristics of a plastic.

date of design | 1999, produced since
manufacturer | Swatch A.G., Biel,
Switzerland

Swatch
Wristwatches

The success of Swatch owes as much to its first
wristwatch—the inexpensive, well-made, tough,
waterproof, precise, analogue model that caught
the public's imagination in 1983—as to its clever
promotional strategies. The firm eventually
became an amalgamation of well-established
names in the Swiss watchmaking industry that
needed an injection of imaginative marketing. The
models here serve as examples of Swatch's continuing efforts toward innovation and cutting-edge
technological advancements.

below | "YSG 109 Cross Talk" woman's watch (25 x 28 x 8mm) has a band and case in PVD-finished 316 L stainless steel. PVD (physical vapor deposition) coatings—extremely thin at possibly 0.05 microns—are a mixture of zirconium nitrite, titanium nitrite, or other metallic-ion combinations, applied under low-vacuum conditions. (However, the combination here is not known.) Highly resistant to wear and discoloration, PVD eliminates the necessity for a clear protective coating of another material.

date of design | 2000, produced from 2001
manufacturer | Swatch A.G., Biel, Switzerland

manufacturer says | "Heavy-metal tricks—these were the sounds of the moment in 1994 when Swatch added metal to its distinguished plastic profile and went from the 'Gent' range to 'Irony' and beyond. Swatch continued to promote its visual identity with its ingenious MIM metal watchcases… now heavier, shinier, and more solid and with a sporty edge. 'Irony' could claim 'Swiss made.'"

Photography © Swatch A.G.

System 180
Furniture

Headed by British designer Chris Smith, System 180 has come up with a modular system that can be assembled into almost any type of furniture. Square, angular, or round forms are possible, only a few examples of which are shown here. The system may also be configured architecturally as staircases and exhibition stands and as shelving units in retail interiors. Smith's role as designer, manufacturer, and marketer illustrates, yet again, how designers are choosing to circumvent the normal practice of seeking out manufacturers to produce their work. (See Smith's "piclip®" picture frame on pp. 112–13.)

facing and below | "System 180," built as large shelving, is essentially composed of tubing assembled with M8 Allen wing bolts and nuts. The tubes are zinc-coated steel but also available in stainless steel. Cutting is by a computer-numeric-controlled (CNC) brake press. The flat surfaces of the desks, stands, and shelving can be solid wood, a laminated material, or glass.

date of design | 1998–2000, produced since
manufacturer | System 180 Ltd, London, UK

manufacturer/designer says | "What can you do with it? Anything!"

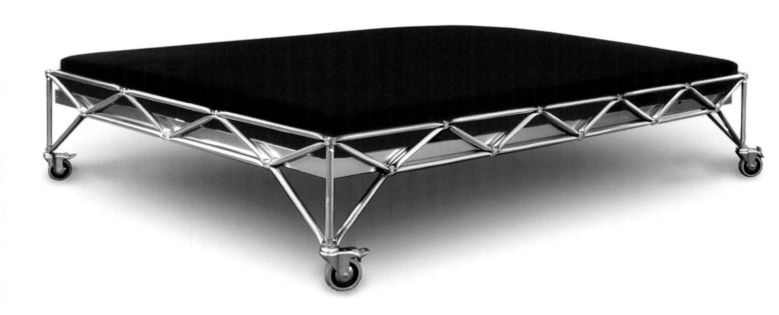

Yuval Tal
Bathroom fixtures

Working in Haifa, Yuval Tal purposefully designed this sink so that the bowl will flex downward when water accumulates. Even though the bowl and table surface are a soft plastic material, it is the steel frame that makes the idea possible. No matter how unusual the concept, it works.

facing | Sink (380 x 910 x 510mm) and shelves (380 x 910 x 380mm) are composed of two materials: stainless steel for the frame and silicone sheets for the surfaces and bowl.

above | The stainless-steel frame is assembled with screws and force-fitted.

date of design | 2001, produced since
manufacturer | The designer

designer says | "A combination of silicon and steel produced a flexible sink that changes according to the amount of water poured into it—a quality created by the use of both soft and hard materials. Unlike a common sink, this one combines easy production, delivery, and installation."

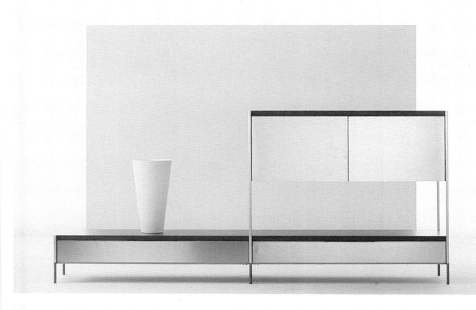

Carlo Tamborini
Modular cabinet system

This storage system by Italian designer Carlo Tamborini permits a wide variety of arrangements to suit individual needs. And, obviously, the steel façade is impressive. Although the cabinets—in whichever combination—are very large pieces, they are not the type of furniture that would be frequently repositioned.

facing and left | "Stilo" (basic container: 520mm deep; top: 35mm thick) is a modular system with supporting frames, uprights, and handles in stainless steel. The façade is also stainless steel, satin-finished. The storage unit boxes are wood (faced with natural oak, iron gray, stained oak, white lacquer, or light gray veneers). Dimensions vary according to built size

date of design | 1999, produced from 2001
manufacturer | Pallucco Italia S.p.A., Castagnole di Paese (TV), Italy

designer says | "The 'Stilo System' is based on the idea of using a light frame to support the horizontal mass of the containers. I wanted to use two different types of materials in order to dismiss the concept of a closed box with a container. For the doors, therefore, we used stainless steel which is usually used for other purposes but, even so, is a rigorous and elegant material."

Photography by Bitetto Chimenti

Ezri Tarazi
Lighting

Since 1996, Ezri Tarazi has been the director of the industrial design department of Jerusalem's Bezalel Academy of Art and Design. He has brought high standards of pedagogy to the century-old institution. An accomplished designer in his own right, his work with candles has brought a new voice to the traditional, soft-spoken lighting fixture. And, on the next page, what appears to be a simple five-branch chandelier is not so simple.

facing and below top row | "Dancing with Light" candlesticks (40 x 120 x 40mm). The bottom screw portion of the candle is an old bulb from which the glass ampule has been removed. The candle mold is a plastic water pipe, the type used for a toilet. The screw portion is attached to the pipe with masking tape; wax is then poured into it. The base is a randomly bent stainless-steel spring wire that is inserted and screwed into the place that would usually house electrical wires.

date of design | 2001, produced since, one-of-a-kind series
manufacturer | The designer

designer says | "The concept behind this object is that the screw portion of a light bulb could be applied to the old world of the candle. Since candles are not extinct but rather very popular, my idea was to design a simple candleholder by upgrading the candle itself with the screw. By using ceramic sockets—traditionally for high-voltage bulbs—I was able to emphasize the fact that no electrical wires are used. So the very thin wires help to achieve the contrast between the heavy ceramic element and the spindly legs. Also, the stainless-steel wires create a wobbly movement."

left | "IUV" candles (70 x 350 x 70mm) contain a vertical piece of mirror-finished and concaved stainless steel (1mm thick) that runs from top to bottom and is attached to a stainless-steel base. Curving and welding machines are used. A metal base is welded to the concaved reflector and connected to the bottom of the casting mold. The wax is then poured into the mold.

date of design | 2001, produced since, one-of-a-kind series
manufacturer | The designer

designer says | "I realized that candles do not have reflectors. Candle light is not fixed like that of an electrical light fixture; it moves down. I had the idea of making a candle reflector that would follow the flame. The wax casting is in three profiles—'I,' 'U,' and 'V,' hence the name. In the 'I' candle, there are two wicks, one on each side of the reflector."

All photography by Yaki Asayag

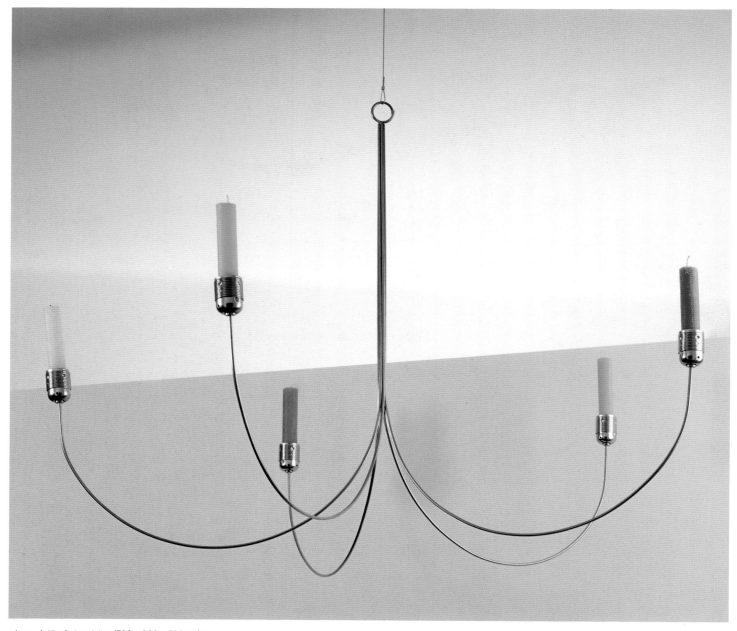

above | "Pen" chandelier (700 x 900 x 700mm) is primarily composed of five stainless-steel rods (dia. 5mm), steel light-bulb screws, and screw candles. The rods are bent and welded with bending, plasma-welding, lathing, and threading machinery.

date of design | 2001, produced since, one-of-a-kind series
manufacturer | The designer

designer says | "In the beginning, chandeliers were made to hold as many candles as possible in order to light big halls. When electric light became available, people still wanted chandeliers and, likewise, electrical fittings in the shape of candles, held in the same vertical position. Now, ironically, the 'Pen' goes backward, using the same form of a traditional chandelier as a reference to the true reason behind the use of candles. But it uses light-bulb screws fitted to wax candles [as on p. 131] as a functional method for holding the candles in place. I have used five steel rods, not tubing, to create wobbling and moving. And I used five candles in five colors as an esoteric reference to the five centers of human energy, a belief of many cultures from Judaism to Hinduism and Buddhism to Islam."

facing | "Flashlight" lamps (160 x 400 x 160mm); two versions: reflector up or reflector down. The parts include mirror-finished, stainless-steel reflectors and base, and springs (around the threaded rod) in handmade spring steel. The rods are screwed to the top reflector. Milling, lathing, and threading machinery is used for the construction.

date of design | 2001, produced since, one-of-a-kind series
manufacturer | The designer

designer says | "The idea for this fixture is based on candles not having reflectors and their light not being fixed like an electrical bulb, but rather moving downward. I wanted the light to be fixed in place and for the bottom of the candle to move upward while melting downward. The hole in the middle of the reflector is a bit more narrow than the candle's diameter. This permits the flame to melt the wax while the candle is being pushed up by three springs. There are two versions. One acts as a downlighter, and the other as an uplighter. The upward-reflected light makes an interesting moving light pattern on the ceiling, like a living fire. And this is the first candle-reflector lamp that has ever been created."

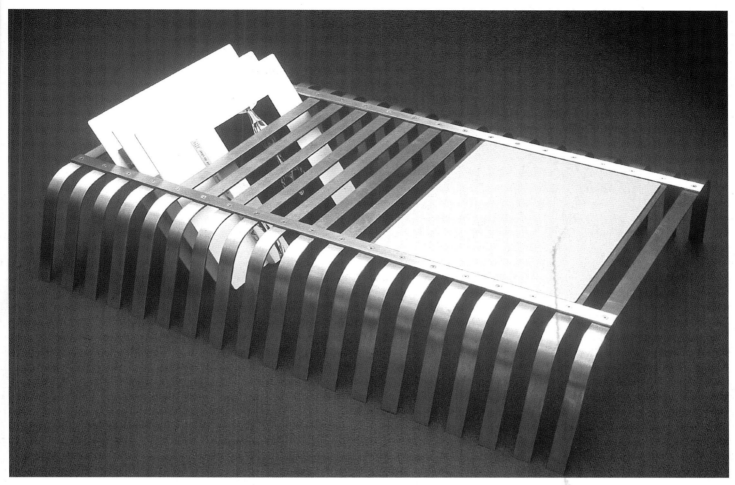

Alexander Taylor
Furniture

Developed for the office or home, this table by London designer Alexander Taylor is constructed of two elements: the cross pieces that serve as both legs and the top surface and linear members, one connected to the other with screws. Various lengths are possible.

top | "Chippy" table (1110 x 215 x 800mm) is brushed AISI 304 stainless steel, with some small hand-polishing. 30 x 5mm strips are bolted with 6mm counters for the Allen-key screws. Computer-numeric-controlled (CNC) machinery is used to cut and bend. The moveable white "tray" offers a smooth surface on which to place objects.

bottom | Close-up view showing the Allen-key-screw connections.

facing | Underside of the "Chippy" table.

date of design | 2001, produced since
manufacturer | Syspal Ltd, Broseley, Shropshire, UK

designer says | "The strength of steel allows for the profiles to be fabricated and retain form; no extra brace or frame is required. The advantage of using brushed stainless steel means that once the fittings are applied there is only a small amount of hand polishing, and the table is complete."

Pekka Toivola
Office furniture

A graduate of Helsinki's University of Industrial Arts and Design, award-winner Pekka Toivola has worked at Martela, a Finnish office-furniture manufacturer, for 15 years. The concept behind his adjustable desk for the firm is that it allows people to sit at it like a traditional low-height desk or to stand at it, facilitating their walking around an office without having to stand up and then sit down, again and again. A motor—operated by a single button, nothing else—is located in the center column to adjust the height. The red rod at the bottom of the column is a foot rest only, and not designed for surface-height adjustment.

| "Promo" adjustable-height work table is offered in two models. Two-section base (shown) adjusts from 700 to 1200mm high; three-section base adjusts from 650 to 1250mm high. The base, feet, cylinders, foot rest, and extension hardware are made of epoxy-painted steel. Work surfaces are laminated or stained beech. An electrical controller adjusts the height; the red tubular-steel foot rest is stationary.

date of design | c. 2000
manufacturer | Martela Oyj, Nummela, Finland

designer says | "[At Martela] it is not my job to come up with all the new ideas. Our sales people, marketers, interior designers, and architects all work together developing solutions. And there is always an abundance of new ideas."

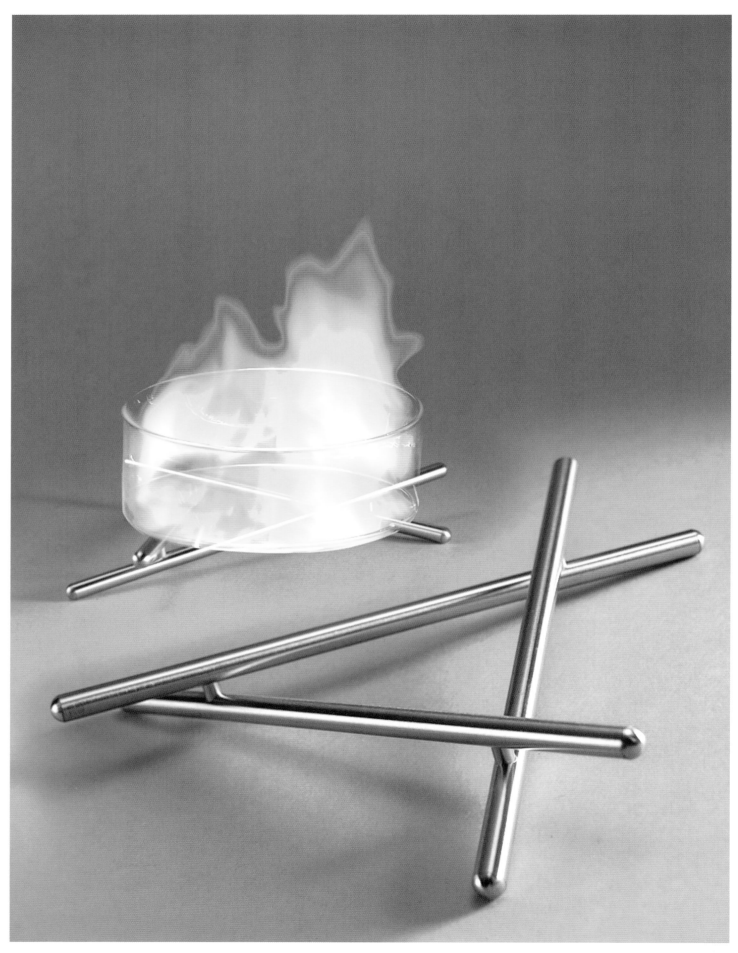

Tools Design
Kitchenware, dinnerware, and accessories

Tommy Larsen prides himself on directing a prize-winning company, founded by him in 1991, as being "100% Danish." The designers he has commissioned include goldsmith Henrik Skjønnemand, Designit (Anders Geert-Jensen and Mikal Jørgensen) of Årnhus, and the team of Jan Egeberg, Morten Thing, and Jan Weilborg of Copenhagen. Claus Jensen and Henrik Holbæk who work from their own studio—Tools Design—also in Copenhagen, have designed the Tommy Larsen items here.

facing | The trivet is composed of three stainless-steel rods separated by welded spacers.

below | The paper towel holder is a two-piece design: a frame in bent stainless-steel wire and an unfinished beechwood dowel.

date of designs | Trivet: 1996, produced since; Paper towel holder: 2000, produced since
manufacturer, pp. 138–41 | Tommy Larsen A|S, Rodelundvej, Denmark.

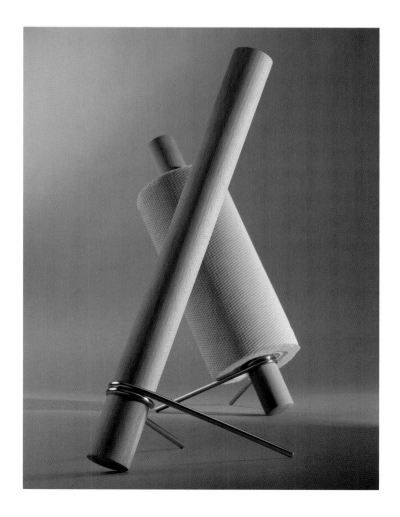

facing top | Key ring could not be simpler—a single bent stainless-steel rod that flexes enough for keys to be slipped onto it but not enough for them to slip off.

facing bottom | Bottle opener is a very thin sheet of stainless steel, the size of a credit card. It received the Formland 97 award.

top | Drying rack includes a beechwood-handle brush and a rack (bent stainless-steel rod and bent base plate). The cloth is placed on the handle for hygienic drying.

bottom | Lemon squeezer is a twisted piece of stainless-steel plate that allows the juice to be extracted directly from the fruit. It received the Formland 95 award.

date of designs | Key ring and bottle opener: 1997, produced since; Drying rack: 1994, produced since; Lemon squeezer: 1995, produced since

---

manufacturer says | "Everything fills a place. Every problem aspires to a solution."

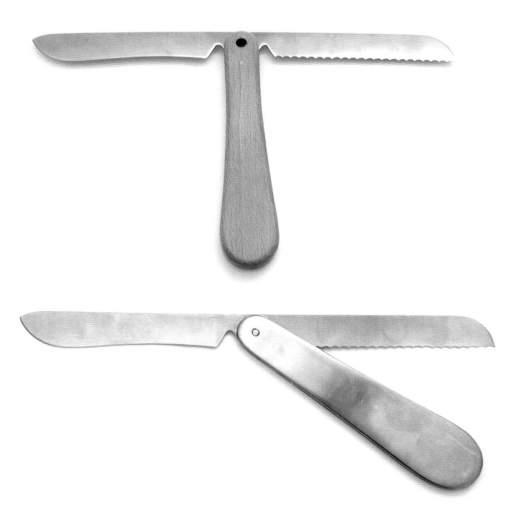

Paolo and Giuseppe Ulian
Kitchen- and dinnerware

A graduate of Florence's ISIA (Istituto Superiore per le Industrie Artistiche) design school, Paolo Ulian has been working in his own studio since 1992. Before then, he assisted Enzo Mari. Preferring to live in Carrara, Italy, as opposed to the design center of Milan, he has become known for his imaginative use of abandoned products and semi-finished industrial scrap. Ulian often collaborates with his brother Giuseppe, but only the kitchen knife shown here was invented by both. Despite the casual developmental drawings shown here, all of Ulian's work is eventually realized as precise technical drawings prior to production.

top | Kitchen knife (320 x 40 x 15mm) by Paolo and Giuseppe Ulian. This product is a double-edged idea. A single, long blade rotates about an axis and is honed on the edge of one half for slicing bread and on the other for slicing salami. When one is in use, the other is housed. Both sections of the handle are the same, united by a screw. The polished blade is laser-cut and machine-sharpened. The handle, available in black or white, is injection-molded polyamide.

bottom left | Black-handled knife opened for salami (top) and for bread (bottom).

bottom right | Elements of a disassembled white-handled model.

date of design | 1999, produced from 2001
manufacturer | Zani & Zani S.p.A., Toscolano (BS), Italy

designer says | "Stainless steel is perfect for the kitchen."
—Paolo Ulian

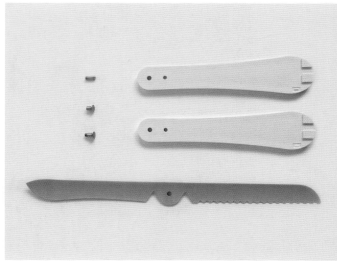

above | Pizza knife (260 x 80 x 20mm) by Paolo Ulian. The utensil comprises just two sections and an axle screw. It is made of machine-bent, high-quality stainless steel, polished or matt-finished.

right | Cardboard maquettes were developed for refinement prior to production.

date of design | 1999, produced from 2001
manufacturer | Zani & Zani S.p.A., Toscolano (BS), Italy

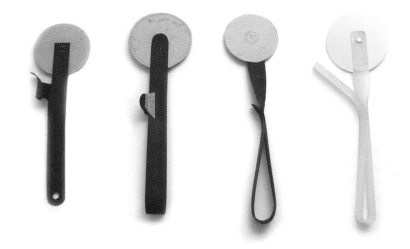

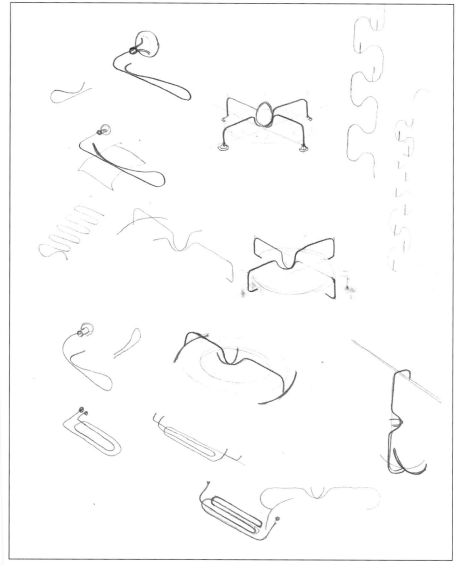

facing | Egg holder (50 x 120 x 280mm) by Paolo Ulian. Three stainless-steel wire pieces were first machine-bent and then welded together.

top | A number of multi-rod configurations were initially considered before the final design was decided upon. The yellow tag on the model second from right says, "OK."

left | Preliminary doodles articulate the quest for simplicity.

date of design | 2000, produced since
manufacturer | The designer

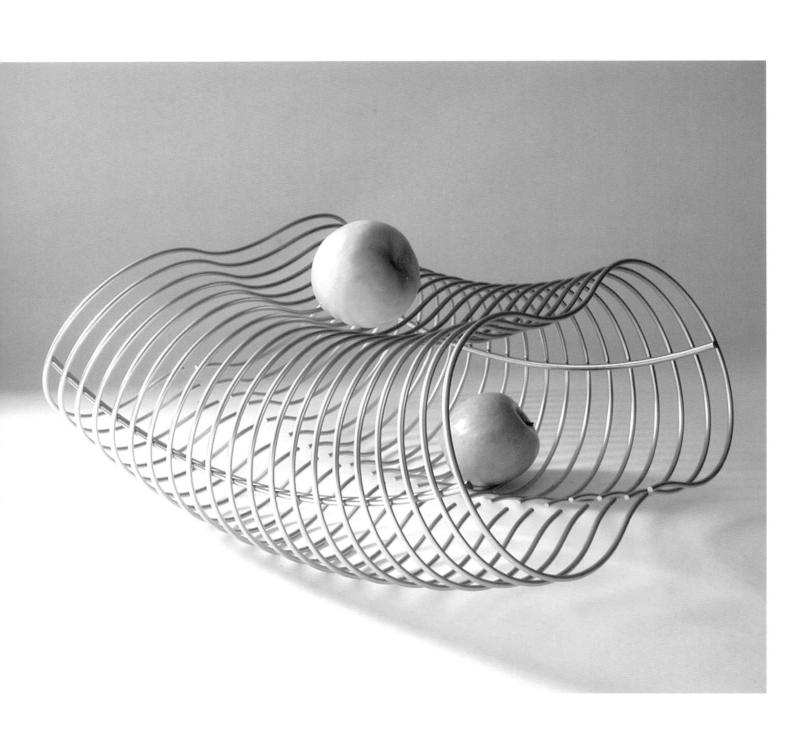

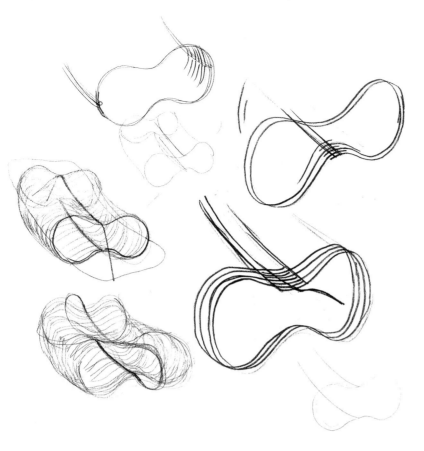

facing | Fruit bowl (400 x 200 x 250mm) by Paolo Ulian. The rod is stainless steel that has been galvanized (zinc-coated), bent by machinery, and welded.

top | A drawing reveals how the concept developed.

middle and bottom | Stages of the initial maquette.

date of design | 2000, produced from 2001
manufacturer | Seccose S.p.A., Preganziol (TV), Italy

All photography by Paolo Ulian

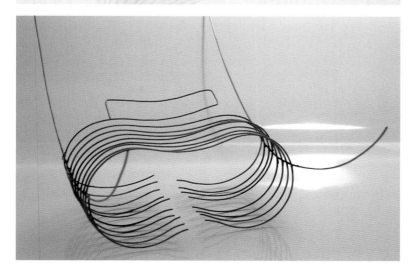

Marcel Wanders
Furniture and bathroom fitting

The best-known object of Dutch designer Marcel Wanders may be his 1996 "Knotted Chair," for which he used a carbon-fiber and aramid rope, woven like macramé and frozen into shape by epoxy. His new work continues this interest in technological processes that are not normally used in furniture-making.

facing | "Flower" chair (650 x 780 x 740mm) is high-gloss chromium-plated steel. The wires of a single flower are first welded, then groups of flowers are welded together in a random pattern. Subsequently, the veil of flowers is welded onto the frame. Molds were used to form the frame and the flowers. After assembly, the chair is plated with high-gloss chromium.

below | Close-up of the welded-wire pattern.

date of design | 2001, produced since
manufacturer | Moooi, Amsterdam, The Netherlands

---

designer says | "All welding is done by hand. After cleaning, the whole chair is covered with a nice, clear, shiny, glossy skin. I like the details to become the structure and the structure to become the surface and the surface to become the chair. In this way, all are connected and become one."

Photography by Maarten van Houten

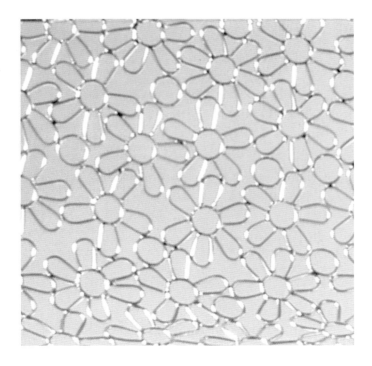

facing | "Henna" table (380 x 450 x 450mm) features a map of Amsterdam that has been etched into the stainless-steel surface—deep into the top, less so on the sides. Laser-cutting and bending machinery produce the form. After assembly with glue, the table is partially lacquered.

above | Close-up detail of the top surface.

date of design | 2000, produced from 2001

manufacturer | Cappellini S.p.A., Arosio (CO), Italy

---

designer says | "Stainless steel makes the drawing timeless and readable for the next 1,000 years. Simple and minimal but decorative and optimal, very rich with surface decoration inspired by henna appliqués, a map of Amsterdam is etched on the table…."

Photography by Maarten van Houten

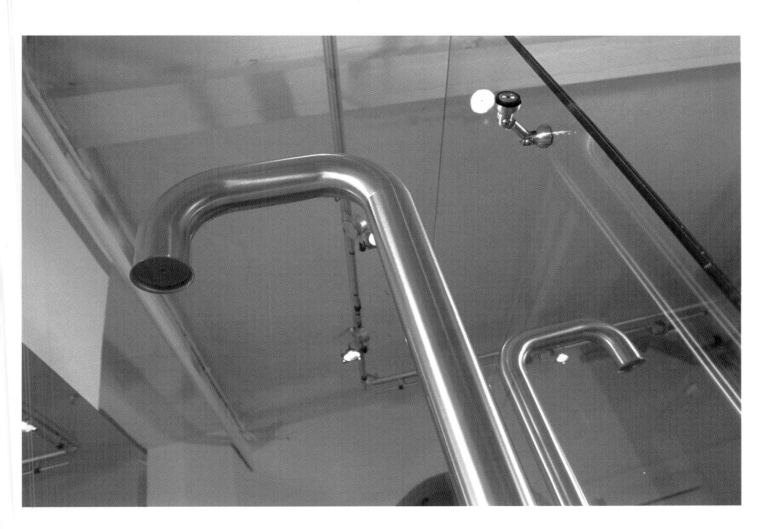

facing and above | "Pipé" shower fitting (2540 x 580 x dia. 90mm) is made of bent stainless-steel tubes.

left | The designer's sketch is more a cartoon than a developmental drawing.

date of design | 2000, produced from 2001
manufacturer | Boffi S.p.A., Lentate sul Seveso (MI), Italy

designer says | "A series of showers, where the water is flowing through thick metal tubes until the water is set free as a waterfall. If we turn a wheel through the air, we receive the radio sounds of a rock band of 20 years ago. If we push a button, we will see a football game played right now in Buenos Aires. If we order our shopping on the 'Net, it will be in the hall at the end of the day. If we want a simple shower, we need a major system of big and small pipes to get the water from the lake to the cleaning facility to the house through the heater and finally through the shower over our naked bodies. Now the 'Pipé' is not a shower; it is a waterfall. It is a thick stream of cleansing, fresh water.... And why in steel? Because real water pipes are made of steel."

Photography courtesy Boffi S.p.A.

Marco Zanuso Jr.
Furniture

Marco Zanuso Jr., the namesake of his late
architect/designer father, created a table with
essentially no secrets upon close examination.
Due to the glass top, the connectors and
assemblage techniques are visible. The cast-
ers insinuate more of an industrial than a
domestic application.

| "Ito" table has a steel base that has been cropped, calendered, drilled, and lathe-worked. Screws and welding were employed in the assembly. Available with a 1300mm dia. round top, 1200 x 1200mm square top, or 1800 x 1200mm oval top (all in heights of 730, 600, or 400mm). The tempered-glass top is 12mm thick.

date of design | 2000, produced since
manufacturer | Driade S.p.A., Fossadello di Caorso (PC), Italy

designer says | "I simply chose to use steel for this table because it is absolutely the best material for rigidity, for weight, and for low manufacturing costs. The top ring, connected to the legs, is made in two sections to facilitate its fitting into a galvanic bath. Moreover, the connection is rather beautiful where the two half-rings (or half ellipses) slightly overlap each other."

Photography by Tom Vack

Carlo Zerbaro and Alessio Bassan
Lighting

Frequent experimentors with the effects of transparency, Carlo Zerbaro and Alessio Bassan have created a lighting fixture composed entirely of stainless-steel parts, with the exception of the opaline glass top protector. This is in contrast to the frames of traditional lamp shades which are normally formed of a lesser grade of wire. Even though the bulbs are visible, the effect is not an unpleasant one because of the density of the wire mesh.

facing | "LEM" is available as hanging (420 x dia. 500mm), floor (1190 or 1700 x dia. 500mm), or table (600 x dia. 400mm) lamps. The shade is stainless-steel mesh. The top diffuser disk is opaline glass. The legs are stainless-steel tubing. The shade armature is stainless-steel wire. Production involved calendering, pressing, and TIG-welding, and the use of laser-cutting and lathing machinery. (For an explanation of TIG welding, see p. 13.)

top | Various parts of the lamp can be identified.

bottom | The lamp and the opaline diffuser seen from above.

date of design | 2001, produced since
manufacturer | &'Costa S.r.l. (Maria Costa), Schio (VI), Italy

designers say | "We love playing with transparent effects, and our research has moved toward new materials that can be used in lighting to offer transparent effects of softness and veiling. Because technology is the main aim of our research, we are aware that stainless steel is a lasting, durable, and easy-to-clean industrial material. Also, we have found that stainless-steel mesh works like a filter—playing a game of hide-and-seek with light—and satisfies all the requirements needed to reach our goal."

Photography by Carlo Zerbaro and Alessio Bassan

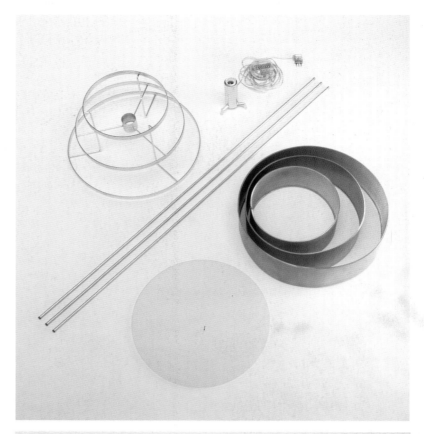

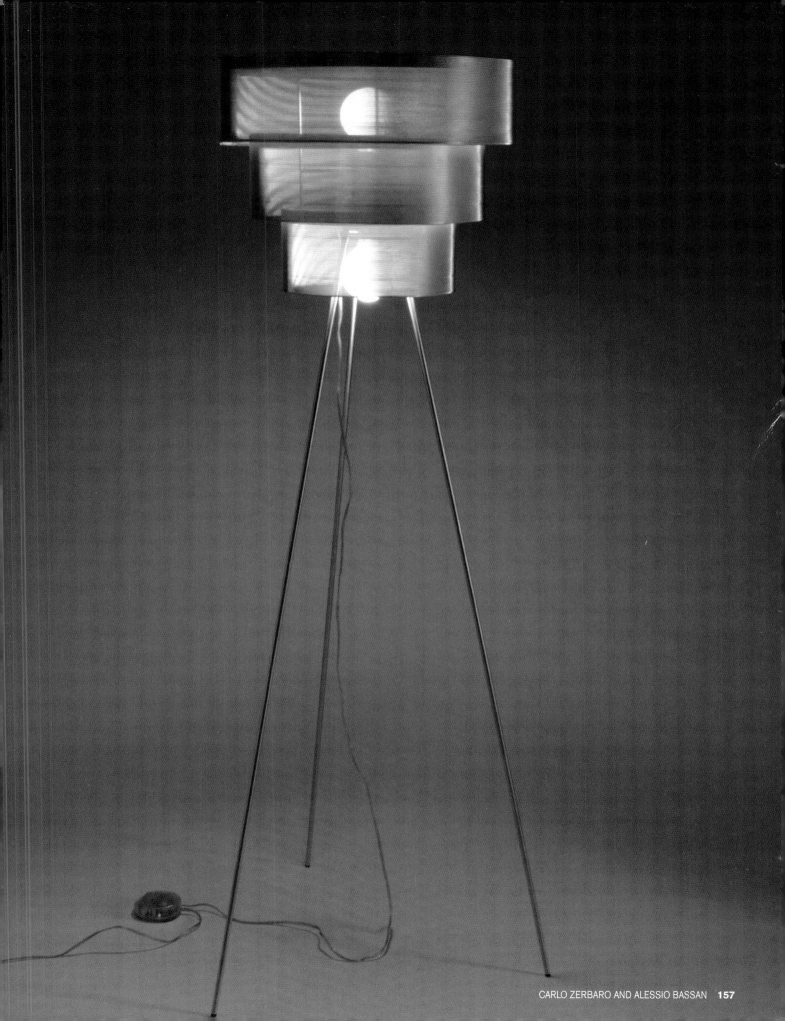